IMAGES
of Aviation

BOWMAN FIELD

On the Cover: While Bowman Field was converted to a military base during World War II, there was still a civilian presence there. Both American and Eastern Airlines continued daily flights in and out of Louisville. Moreover, Louisville Flying Service continued to serve civilian aviation. (Edward Peck.)

IMAGES
of Aviation

BOWMAN FIELD

Charles W. Arrington

ARCADIA
PUBLISHING

ISBN 978-1-4671-2699-1

Published by Arcadia Publishing
Charleston, South Carolina

Printed in the United States of America

Library of Congress Control Number: 2017934567

For all general information, please contact Arcadia Publishing:
Telephone 843-853-2070
Fax 843-853-0044
E-mail sales@arcadiapublishing.com
For customer service and orders:
Toll-Free 1-888-313-2665

Visit us on the Internet at www.arcadiapublishing.com

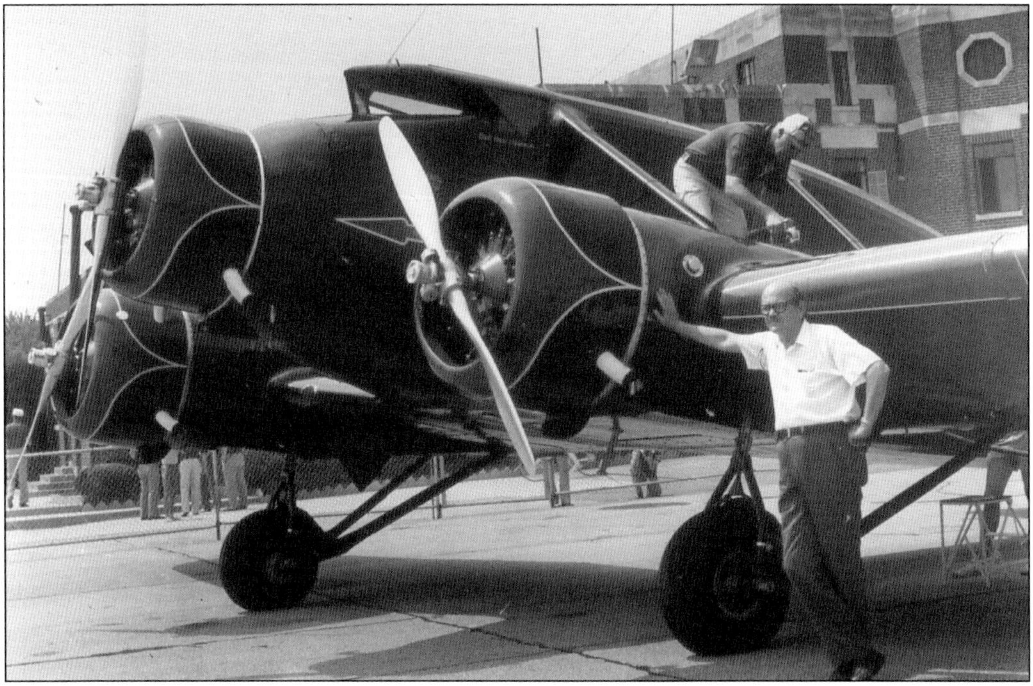

This book is for you, Ed Peck. The path that you paved made it possible.

Contents

Acknowledgments 6
Introduction 7
1. From Cornfield to Airfield 11
2. Building Bowman Field 25
3. World War II 53
4. Bowman Field Takes Flight 77
5. Air Shows, Fly-Ins, and Festivals 101
6. Looking Back, Looking Forward 109

ACKNOWLEDGMENTS

First and foremost, this book would not be possible without the significant Bowman Field research done by the late Merritt Edward "Ed" Peck (1925–1998). Peck was Kentucky's premiere aviation historian. Much of his photograph collection was obtained from original sources far and wide. He devoted much of his later life to Bowman Field research, and through his efforts and treasure, he got three historic buildings at Bowman Field listed in the National Register of Historic Places.

Another important resource was the late Chester Lamppin. Lamppin was a member of the Bowman Field reserve squadron and was there photographing events, often next to newspaper photographers. His sharp memory was invaluable in piecing together the story of Bowman Field. Acknowledgement must also be made to Edward O. Taylor for his contribution to researching Bowman Field history. And a special salute to Bob Dearing for making the Bowman Field of the 1940s and 1950s come alive once again. William Russell Beeler must be mentioned as the man who always carried a camera.

Sincere appreciation is extended to the many photographers and photographic sources that made this book possible. They include the University of Louisville Photo Archives, the Louisville *Courier-Journal* newspaper, and Regional Airport Authority (RAA). Photographs were also obtained from the US Army Corps of Engineers (COE). Among the many individual photographers, a thank-you is extended to Gary Chambers, Charles Thornton, and Charles Keeling via his daughter Barbara.

A word of thanks is extended to the many individuals who contributed to the book in one way or another. They include Joe Hardesty of the Louisville Free Public Library (LFPL), Robert L. Burns, Robert Troutman, Bill Stebbins and his wife Mary Joe, the Dick Mulloy family, Wil Paris, Angela Calloway Washington, Rick Tabb, Wilma Shackleford Ford, Ethel Guffey Simpson, Tom Hubbuch, and Becky Carnes.

Insuring that this book became a reality, appreciation is extended to Rick Tabb for his work, guidance, and advice, along with his pushing and shoving to get the job done. Lastly, thanks to Liz Gurley of Arcadia Publishing for making the road traveled toward an Arcadia book a smooth one.

Introduction

Bowman Field is Louisville's iconic century-old airport, where things have changed but not changed. The airlines and military are long gone, but the age-old magic of an airport lives on. The magnificent terminal building beckons one to stop and look and sit in its shade to watch airplanes. Like the mythological siren of old, Bowman Field has attracted those who dream of flight, and hopefully, it will for another 100 years. Bowman Field is Kentucky's oldest continually operated airport and one of the oldest in the United States.

A number of Louisvillians became military pilots during World War I, and some served overseas. They developed a sense of adventure and a love for flying and wanted to continue doing it after the war. One was Robert Henry "Bob" Gast, who joined the British Royal Flying Corps and flew several bomber missions over Germany late in the war. He returned to Louisville in March 1919 and looked for local flying opportunities, although he had to depend on employment as an auto mechanic.

By 1920, another Louisvillian, Abram Hite Bowman, began to visualize the importance of airplanes and the need for his city to reach into the future with an airport. Bowman's business was trucking, and it was perhaps through Gast's employment as an auto mechanic and Bowman's trucks that the two met and formed the Bowman-Gast Aero Company on May 8, 1920, and opened Bowman Field. Bowman's influence and financial strength along with Gast's piloting skills would get the new airport off to a start. One of the first purchases was a World War I Canadian-built Curtiss Jenny, better known as a "Canuck."

The land selected for Bowman Field had the advantage of being east of Louisville, with excellent surface transportation via Taylorsville Road and a parallel interurban rail car line that extended to Jeffersontown. The location was also free of the industrial air pollution from Louisville. The downside of the airport land was that it was owned by a German national named Waldemar von Zediwitz. Von Zediwitz was a descendant of the original owners of the land that Bowman Field mainly occupied, the Breckinridge-Caldwell family. Because he was an officer in the German army during World War I, his land holdings in the United States were confiscated during the war by provisions of the Alien Property Act. During 1920, Abram Bowman rented the land from farmers who had an already established lease on the property. After the war, von Zediwitz wanted his land returned, which resulted in court action and legal procedures that lasted until 1928.

Unfortunately, it was not profitable to operate an airport in the early 1920s. There were no airlines or the ability to fly heavy air cargo at that time. Sightseeing rides and servicing passing barnstormers was about the extent of Bowman Field's early business. Bob Gast left in 1921 in search of more lucrative flying opportunities. In a short time, Abram Bowman had a new partner, Wallace Sidney Park, another Louisvillian. Park had just been commissioned a pilot in the US Army Air Service and put on inactive reserve status. His specialty in the military was aerial photography, and this brought a new dimension to the reincorporated Bowman-Park Aerial Company. Bowman, Gast, and Park have long been considered the founders of Bowman Field.

During 1922, word came that the Army Air Service would abandon Godman Field at Camp Knox, today known as Fort Knox, as a result of military downsizing following World War I. Taking the initiative, Louisville mayor Huston Quin and local congressman Charles Ogden made arrangements for Ogden to meet in Washington, DC, with the War Department to offer the transfer of Bowman Field to the Army Air Service and establish an Air Service Reserve squadron at the field. Louisville would provide the airfield rent free for five years and contribute money in support. Later, a delegation that included Mayor Quin and Abram Bowman traveled to Washington to discuss details. The proposal was approved on October 20, 1922, and provided for the establishment of the 465th Pursuit Squadron Reserve, US Army Air Service, and the transfer of two hangars and three small buildings from Godman Field to Bowman Field at the city's expense. Bowman Field now had the military designation Bowman Army Airdrome.

On October 26, 1922, a meeting was held in Mayor Quin's office to discuss the new Bowman Army Airdrome. Of 25 attending, 18 were former military aviators with service during World War I, and the others were local aviation supporters. They organized the Aero Club of Kentucky and promised to be actively involved in the airport. Capt. Joseph D. Peeler was elected president, Lt. Albert Woody vice president, and Lt. Weisley Keister secretary/treasurer. Abram Bowman and Sidney Park also attended the meeting. In time, both Peeler and Woody would be reserve squadron commanders, and Bowman would become president. The club is still active at Bowman Field to this day as the Aero Club of Louisville.

A number of positive factors came about with this transfer. Perhaps most important, Abram Bowman was no longer responsible for the airport and its operation. Furthermore, the first steps of making Bowman Field a municipal airport had begun. Abram Bowman would work tirelessly for the rest of his life in support of Bowman Field. Through his position in business and the community, he expressed his belief that Louisville would not be a successful city in the modern world without a municipal airport and inclusion on the national air map that would bring progress.

Perhaps the most interesting provision of the newly formed 465th Pursuit Squadron Reserve at Bowman Field was that it was encouraged to work closely with civilian aviation and allow it to sell gasoline, oil, and parts as well as hangar civilian airplanes. The first aircraft assigned were Curtiss JN Jennies for training. By the late 1920s, it was decided that the 465th did not fit the mold of a pursuit squadron, which in modern terms is a fighter squadron, so it was changed to the 325th Observation Squadron Reserve and re-equipped with observation aircraft with the capability of aerial photography.

Perhaps the most exciting event in Bowman Field's short life was the visit by Charles Lindbergh in August 1927. He was the hero of the day after his lone flight across the Atlantic to Paris earlier in May. Lindbergh landed at Bowman Field on August 8, flying his famous *Spirit of St. Louis*. He was accompanied by support pilot and friend Philip "Red" Love, who was flying a single-engine Fairchild. It is said that 100,000 people lined Broadway to view Lindbergh's motorcade that took him downtown to the Brown Hotel. At a dinner in his honor, Lindbergh gave a speech promoting local aviation in Louisville. He said, "You have made improvements here and the field [Bowman Field] will bring to Louisville air lines and many advantages of aviation."

The year 1927 also saw the legal affair with von Zediwitz over land ownership come to a head. Although von Zediwitz lost his court cases, Pres. Calvin Coolidge restored all property to foreign aliens who had it confiscated during World War I. Bowman Field now had to be paid for or lost. A local bond issue was proposed and presented to Louisville voters that year; it would allow the city, through the parks department, to raise $750,000 to buy the von Zediwitz land and keep Bowman Field along with creating Seneca Park. Thanks to the response of Lindbergh's visit and speech in August, the bond issue was approved. Bonds were sold in 1928, and that year, the Kentucky state legislature enabled municipal airports in Kentucky. Bowman Field became Kentucky's first municipal airport and was to be governed by the Louisville and Jefferson County Air Board.

The biggest event in 1928 for Bowman Field was the arrival of air mail. The US Post Office was an early believer in the airplane and started air mail flights in 1918 using the military. In 1925, the Kelly Act shifted air mail from the military to civilians with the establishment of Contract Air Mail (CAM) routes. Since pilot safety was of prime concern, air mail routes could not be opened until night lights and auxiliary airfields for emergencies were in place. Bowman Field's first air mail was designated CAM Route 16 and followed a path from Cleveland to Akron, Columbus, Dayton, and Cincinnati before ending in Louisville. Between Louisville and Cincinnati, auxiliary or emergency airfields were established and nighttime beacon lights mounted on towers every 15 miles. The first air mail flight to Louisville arrived on August 1, 1928.

With the beginning of air mail and the birth of airlines across the nation, airports now had a purpose but needed investment. Bowman Field's first out-of-state commitment came in 1929 with the arrival of New York–based Curtiss Flying Service. After its merger with the Wright Aeronautical Company that year, Curtiss became the largest producer of airplanes in the country. To promote business and teach customers to fly, Curtiss established nationwide locations, including Bowman

Field. A building that provided offices, classrooms, and hangar space was dedicated by Curtiss Flying Service at Bowman Field on July 5, 1929. Curtiss Flying Service was a major boost to Bowman Field, as it brought with it a national name with a familiar product: Curtiss airplanes.

The first major national airline offering passenger service to Bowman Field was American Airways, which inaugurated operations on March 2, 1931, with a mail-only flight between Louisville and Nashville. Passenger operations began on April 30 with a route that connected Cincinnati to Atlanta via Louisville. The launch included two airplanes in an inaugural flight carrying only company and community leaders. Public flights started the next day connecting Cincinnati, Louisville, Nashville, Chattanooga, and Atlanta. Eastern Air Lines followed in 1934, starting with air mail service on June 2. Passenger flights began on December 20 on a north-south route that linked Louisville with Chicago, Atlanta, Jacksonville, and Miami.

The Curtiss-Wright Aircraft Corporation, hard hit by the Great Depression, took steps to downsize. One victim was its Louisville location. Fortunately for Bowman Field, John C. Bennett was now the manager of Louisville's Curtiss Flying Service. Bennett took immediate steps to keep the Louisville operation open. He traveled to New York City and successfully made arrangements to obtain ownership. As a result, he renamed the business Louisville Flying Service. Next, he secured a bank loan to buy the equipment left by Curtiss-Wright. The date of the new incorporation was June 2, 1932. In partnership with Bennett was aircraft mechanic Frank Holzaphel, whose high standards were always the envy of the airport. This dynamic duo kept the business open throughout the Depression years to serve Louisville's aviation needs.

As Bowman Field's two airlines added flights, it was evident that a larger administration building was needed. In 1936, it was decided to triple the size of the building, which opened in 1929, as a Depression-era Work Progress Administration (WPA) project. The new design featured mirror-image east and west sides and added a large center section. The building included a restaurant and a large, glassed-in aircraft traffic control tower on top. The dominant room was a large lobby with airline ticket windows and seating for waiting passengers. This beautiful building stands today almost untouched by the hands of time.

With the 1930s, Bowman Field's 325th Observation Squadron Reserve was in need of new facilities. It was still using the two small metal hangars from a decade earlier. In 1932, news came from the War Department that two new attached hangars would be completed. Constructed of brick, steel, and concrete, the hangars were publicly dedicated on Saturday, October 22, 1932, to much fanfare. The 325th Observation Squadron Reserve participated with a nine-plane aerial review. The two hangars are still at Bowman Field today and are currently used by Central American Airways.

When Germany attacked Poland on September 1, 1939, starting World War II, war seemed a long way from home. By 1940, expansion of the US military was underway, and Bowman Field was not to be left out. Information came on October 3, 1940, that Bowman Field was to be made larger and become a military airfield. The new Bowman Army Air Base would host the 46th Bomb Group and its four squadrons of Douglas A-20 attack bombers, which arrived on May 18, 1941. The new air base would be on the east side of the field. After the United States' entry into World War II with the attack on Pearl Harbor, it was decided to change the role of the new Bowman Army Air Base from bombers to troop transport. As the bombers departed, the 1st Troop Carrier Command came to Bowman Field. Its mission was to train crews and build flying proficiency to prepare for combat.

Closely connected to troop transport was the School of Air Evacuation. This school set Bowman Field apart from other stateside training bases, as it was the only one in the entire Army Air Forces from 1942 to 1944. It brought together doctors, nurses, and medical technicians and trained them in air evacuation skills for use on the front lines. It was at Bowman Field that the concept of the flight nurse was developed with the idea of saving lives. Many more lives could be saved if wounded could be quickly airlifted to a hospital behind the lines.

Another role for Bowman Field was the training of combat glider pilots. Starting in March 1943, already-trained glider pilots were brought to the Bowman Army Air Base for advanced instruction. Originally called the Glider Pilot Combat Training Unit but later named the Glider Crew Training Center, this operation at Bowman Field lasted one year. As originally conceived, glider pilots were to become infantry soldiers after they landed in a combat zone, so it was important to train them as riflemen. Weapons training was usually done at Fort Knox. Few gliders were actually assigned to Bowman Field; small Grasshopper airplanes, mainly Taylorcraft L-2s and Aeronca L-3s, were used to gain flying hours. Perhaps the best known glider pilot at Bowman was child actor Jackie Coogan. Coogan liked to go off base for lunch; he particularly enjoyed Kaelin's Restaurant for its cheeseburgers.

Into 1944, when victory in the war with Germany and Japan seemed possible, the US military began to reduce training at bases across the country, including at Bowman Field. First to leave was the combat glider school in March 1944. Next, the School of Air Evacuation was reassigned to Randolph Field, Texas, that October. Finally, all training at the Bowman Army Air Base ended on November 1, 1944, and Bowman Field was reassigned to the Personnel Distribution Command to become a convalescent center for returning troops. This continued beyond the end of the war.

After the war, Bowman Field slowly worked its way back to a totally civilian airport. One of the provisions of the World War II GI Bill approved by Congress was that it offered flight training to veterans. Entrepreneurs at Bowman Field organized training operations to meet this need. In time, the field had 22 flight schools on the premises. Calloway Flying Service, operated by Julius Calloway, a former Tuskegee-trained fighter pilot, ran a school for African American veterans. Unfortunately, the GI Bill training era quickly ended. By the late 1940s, it was all over, due in part to the lack of interest by veterans and an abundance of government regulations.

The 1950s brought lots of changes to Bowman Field, introducing a new beginning for what was now Louisville's general aviation airport after the airlines moved to the new Standiford Field on November 15, 1947. The 1950s also brought a surge of interest in flying, and Bowman Field was ready. On the east side of the field, Kentucky Flying Service, owned by Dick Mulloy, became one of the most popular flying locations at Bowman Field. Kentucky Flying Service created lots of happy fliers and lasting memories. Louisville Flying Service, still with John Bennett and Frank Holzaphel, did the same but also catered to corporate aircraft owners. Central American Airways, led by Wil Paris, specialized in charter operations and had several large airplanes to carry cargo and passengers. Kentucky Air Transport, owned by J.B. Wathen, was the smallest of the four, and during the 1960s, changed its name to Cardinal Aviation.

The flying boom that came to Bowman Field during the 1950s was unparalleled. In 1957, with 175,266 local flight movements, Bowman Field was named the busiest local traffic airport in the United States by the Federal Aviation Administration (FAA). Some of the reasons given for this achievement included being close to downtown, ease of travel to the airport, being close to residential areas, and competitive prices among the flying services.

Unfortunately, the boom could not last, and a period of decline set in at Bowman Field. Reasons for this included the rising cost of gasoline and insurance and the negative impact of aviation during the Vietnam War. Furthermore, in a quickly changing world, people had so many other things to do. As popular airport flying services closed because of owner retirement or lack of business, outside interests often took over. Although efficient, they sometimes had difficulty being profitable during the downturn.

As always, things have a way of coming around, and 21st-century Bowman Field is in a period of renaissance. A new enthusiasm and interest in flying is evident on a daily basis just by watching air traffic in and out of the airport. Fueling this is the immense popularity of homebuilt aircraft and the numerous new brands on the market offering airplanes made of composite materials.

A hundred years ago, pilots flying at Bowman Field flew by the "seat of their pants," as the old expression goes. They followed rivers, roads, and the "iron compass" (railroad tracks) for direction. Today, Bowman Field is a bit more organized, with runways, air traffic control, and devices for navigation. Still, it is hoped that the same sense of adventure and thrill of flight that attracted the early aviators will continue. It is truly hoped that the spirit of flight that has always been at Bowman Field will be there for the next 100 years.

One

From Cornfield to Airfield

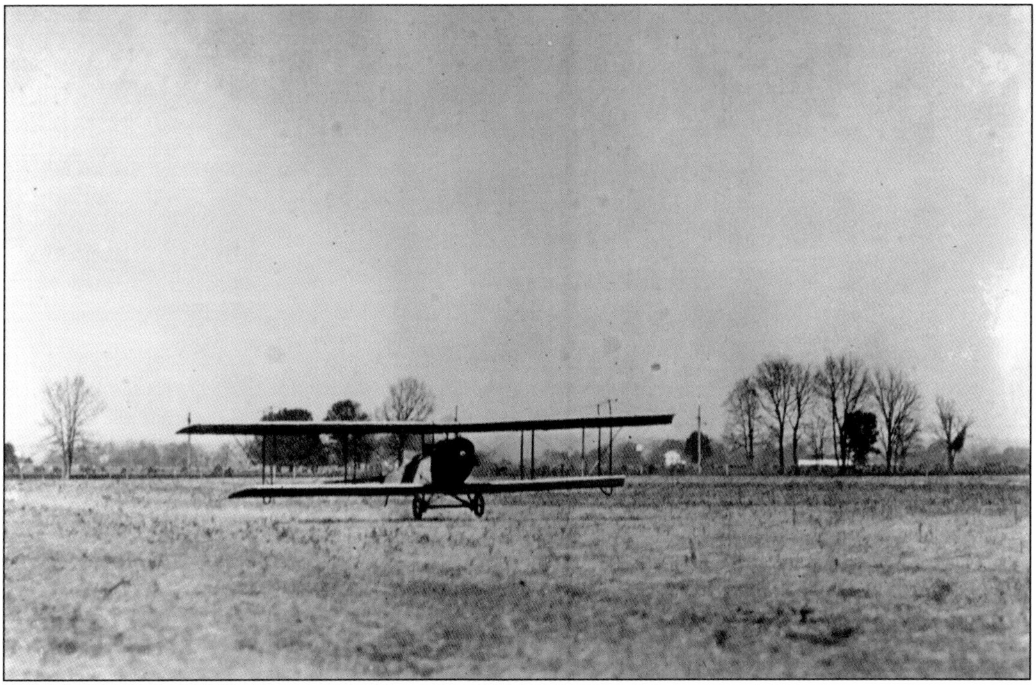

The Bowman Field of 1920 would hardly be recognized today. It was originally 50 acres of corn field and cow pasture. There were no runways or modern buildings. Airplane pilots would check the wind and take off in the appropriate direction. The most common airplane at this time, as seen here, was the Curtiss JN-4 Jenny. It had a 90-horsepower engine and carried one pilot and one passenger. It had no brakes and needed someone to flip the propeller to start the engine. From this simple start, Bowman Field evolved into the modern airport that it is today. (RAA.)

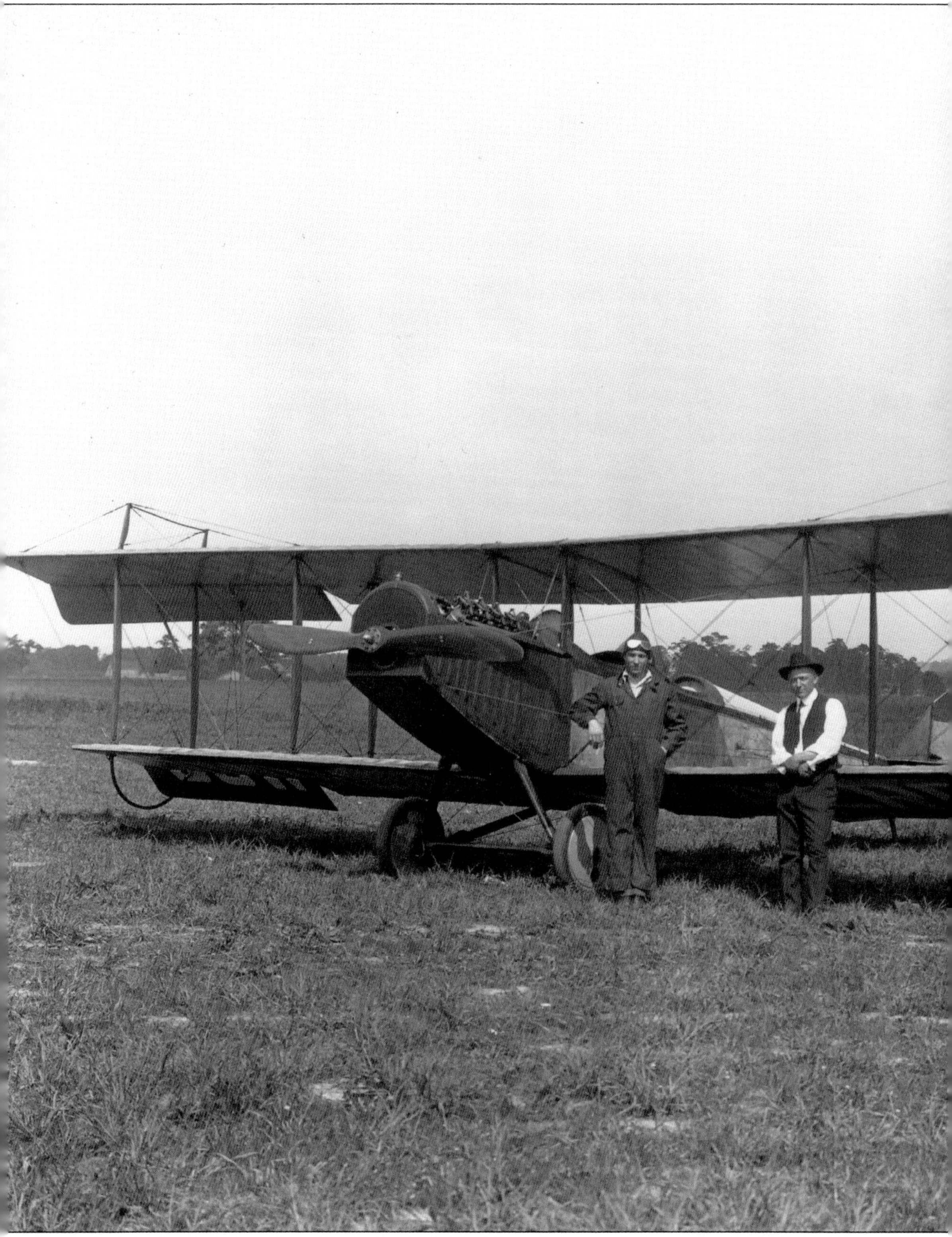

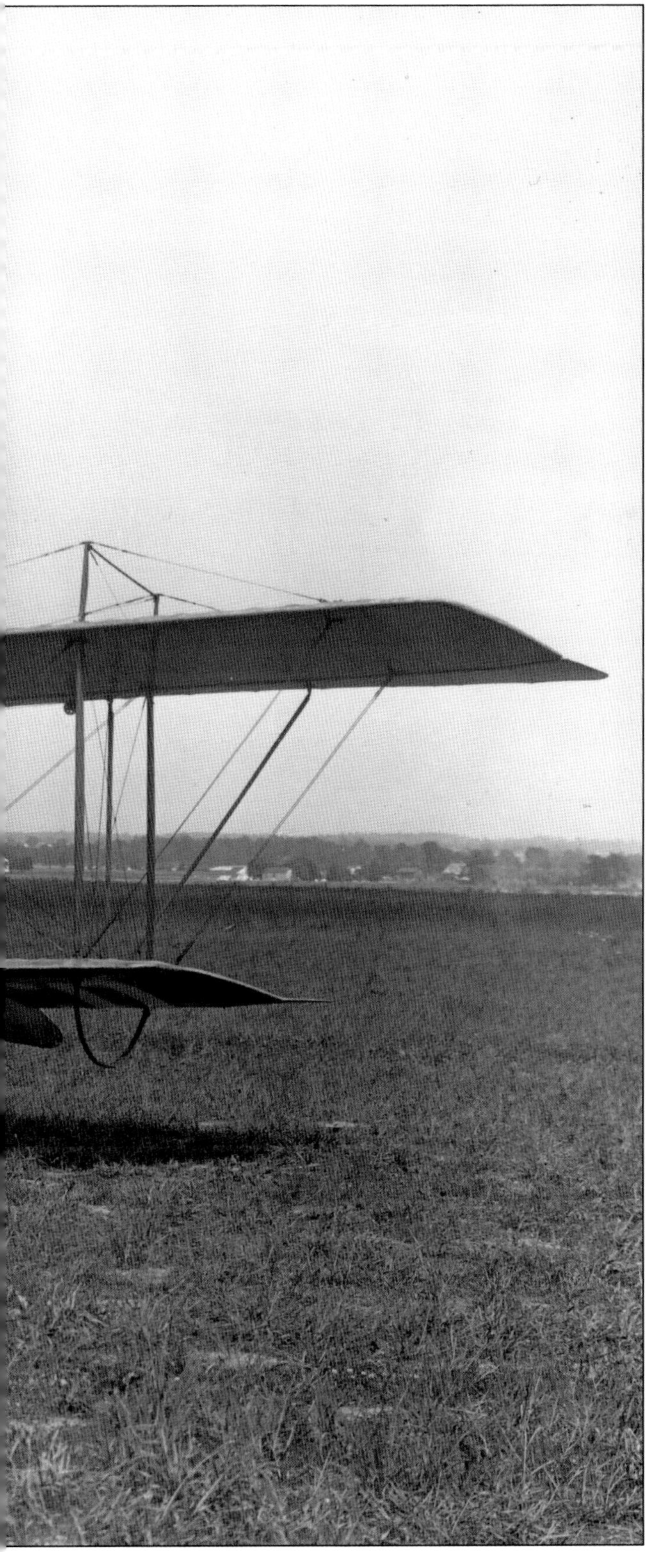

Perhaps the most iconic of all Bowman Field photographs is this one from 1920. It shows Bob Gast and Abram Hite Bowman standing in front of their first airplane. The airplane is a Canadian-built Curtiss Jenny known as the Canuck. This plane started the Bowman-Gast Aero Company, which was incorporated on May 8, 1920. Flying duties were performed by Bob Gast, who had flying experience from World War I. Local sightseeing flights took up most of his flying time, although he would sometimes reach out for the unusual, such as the time he used the airplane to pick up money from a Shelbyville, Kentucky, bank and deposit it on the roof of another bank in downtown Louisville. An aerial stunt indeed, but it did let Louisvillians know that there was a new airplane in town. (Edward Peck.)

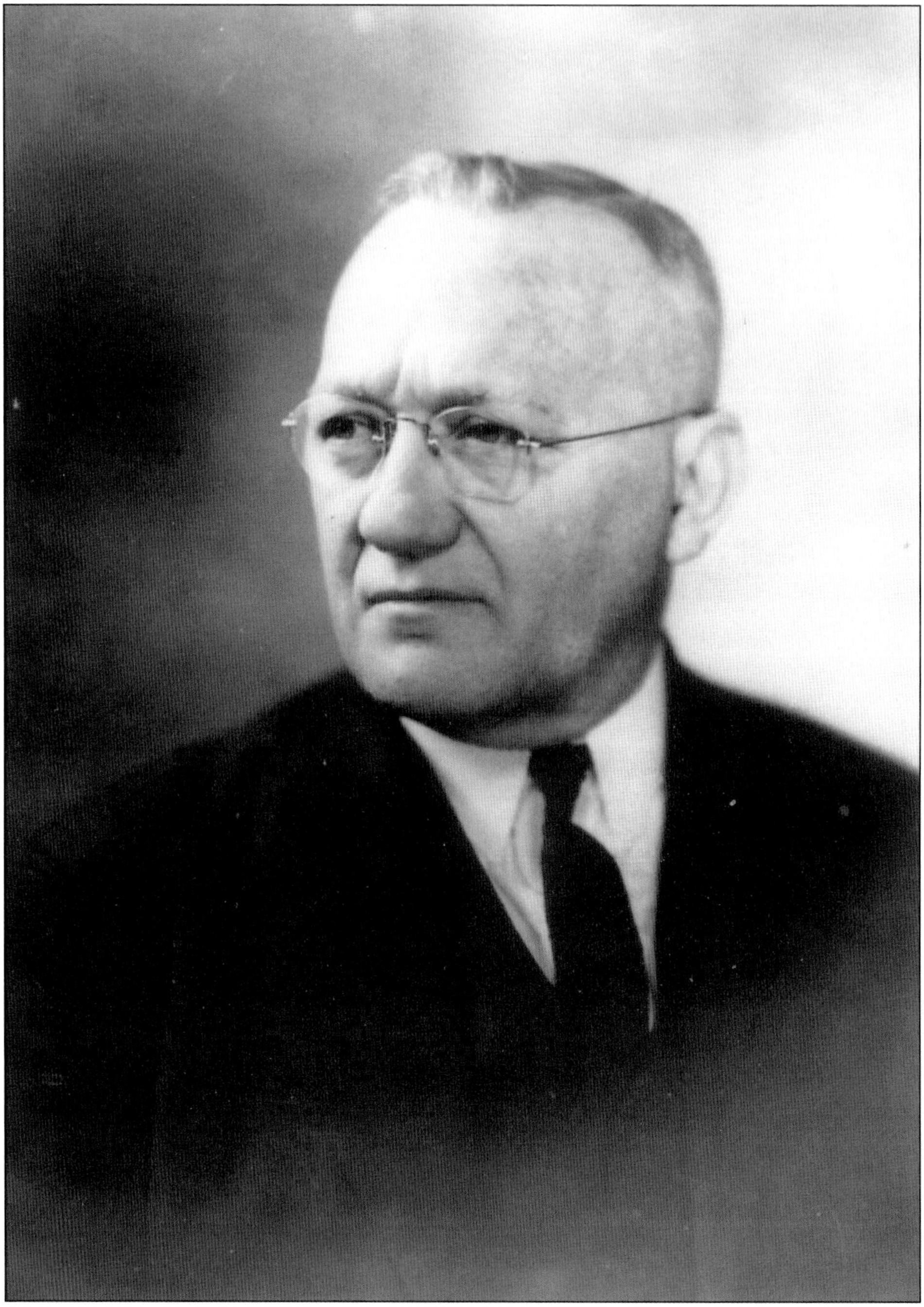

Abram Hite Bowman lived from 1875 to 1943. Known as the "Father of Louisville Aviation," Bowman could see early on the importance of Louisville having a modern airport and taking its part in the progress of aviation. He devoted the better part of his life to this goal. Although not a pilot, he owned a trucking company and was a bank president. He looked forward to the day when the airplane would play a major role in commerce and be an equal partner in the nation's transportation network. The *Courier-Journal* noted at his passing that Bowman Field "bears his name because literally, by the power of persistence and persuasion, he put it there." (LFPL.)

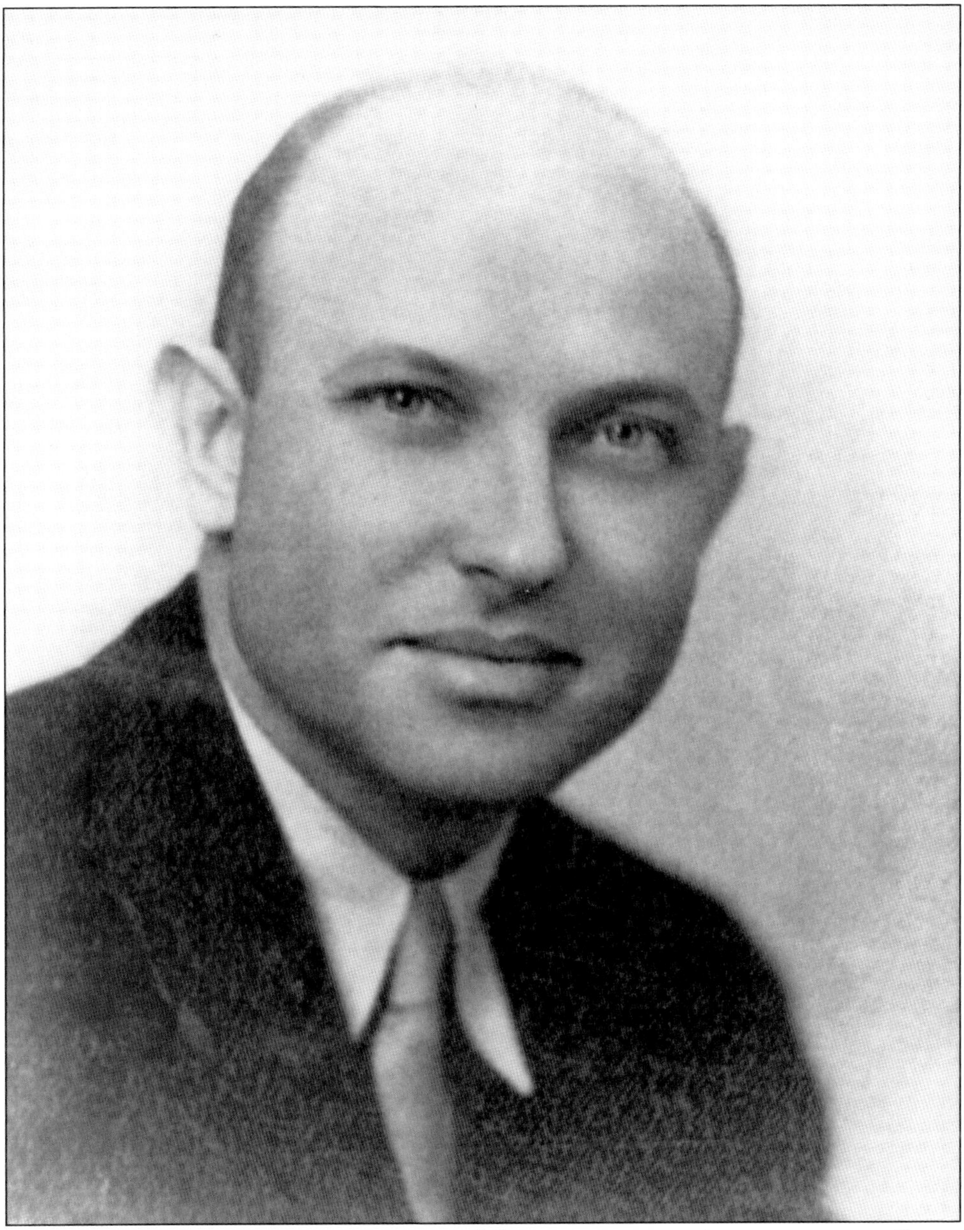

Robert Henry Gast lived from 1896 to 1934. If there was anyone who was meant to fly, it was Bob Gast. His love of flight and sense of adventure took him beyond his partnership with Abram Bowman and the founding of Bowman Field. While in Louisville, he was a pilot with the 465th Pursuit Squadron Reserve and a member of the Aero Club of Kentucky. Away from Louisville, he had his most interesting flying experiences. In 1927, Gast was hired by the US Department of Commerce to be a national inspector. In 1929, he was selected by the *Chicago Herald Tribune* to fly their Sikorsky amphibian airplane from Chicago to Berlin, Germany. The attempt failed when the airplane was crushed by ice in the upper Hudson Bay of Canada. He piloted for China National Aviation Corporation and was killed on April 10, 1934, when his airplane went down in the South China Sea during fog. A memorial was later dedicated to him at Bowman Field, and a street at the airport is named for him. (Edward Peck.)

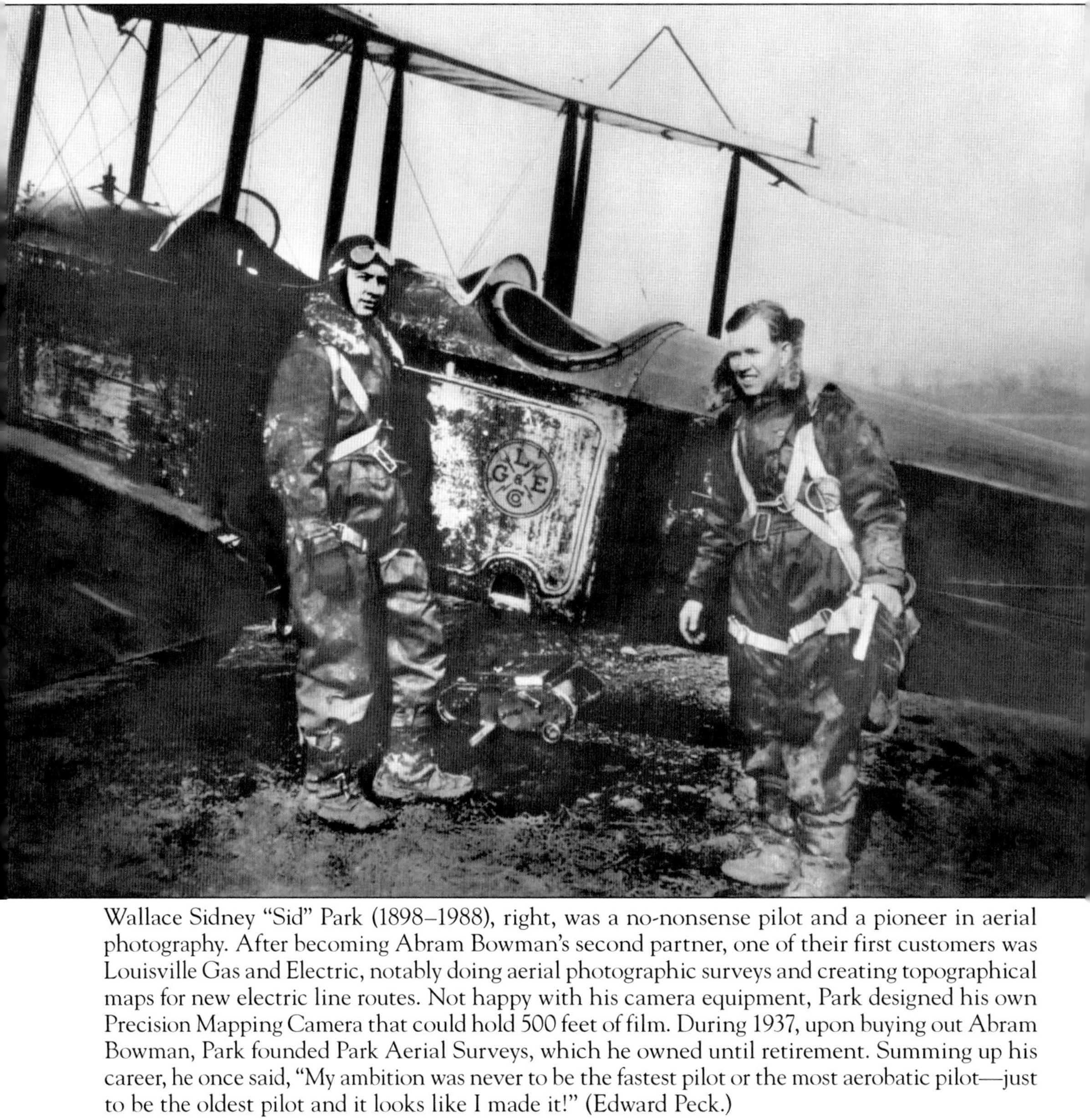

Wallace Sidney "Sid" Park (1898–1988), right, was a no-nonsense pilot and a pioneer in aerial photography. After becoming Abram Bowman's second partner, one of their first customers was Louisville Gas and Electric, notably doing aerial photographic surveys and creating topographical maps for new electric line routes. Not happy with his camera equipment, Park designed his own Precision Mapping Camera that could hold 500 feet of film. During 1937, upon buying out Abram Bowman, Park founded Park Aerial Surveys, which he owned until retirement. Summing up his career, he once said, "My ambition was never to be the fastest pilot or the most aerobatic pilot—just to be the oldest pilot and it looks like I made it!" (Edward Peck.)

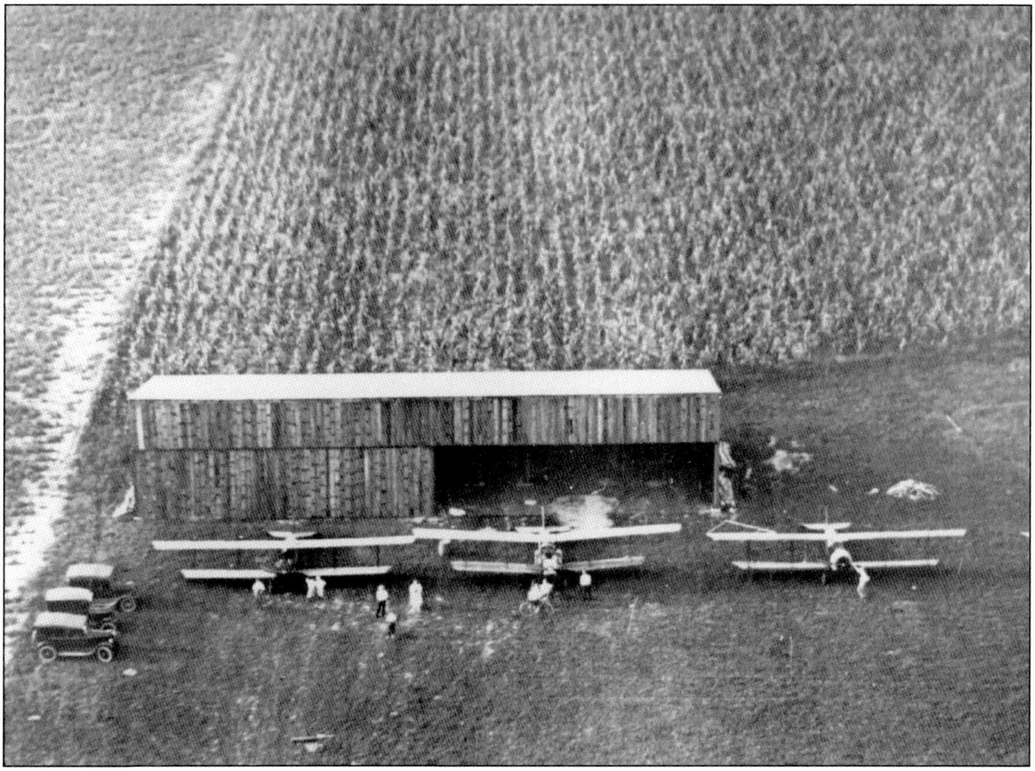

The first hangar at Bowman Field was completed on July 15, 1921. It was located next to a cornfield on the airfield's southwest side. This hangar was constructed using lumber that came from Camp Zachery Taylor as the camp was being demolished following World War I. It was knocked down during a storm on May 13, 1922. Damaged airplanes inside were repaired. Airplanes that appear in these aerial views are believed to belong to an air circus that was passing through Louisville. (Both, Edward Peck.)

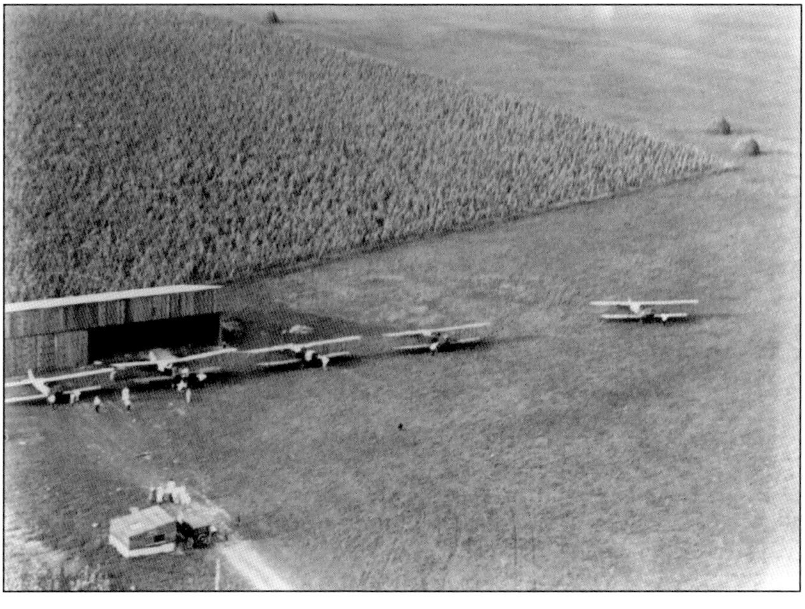

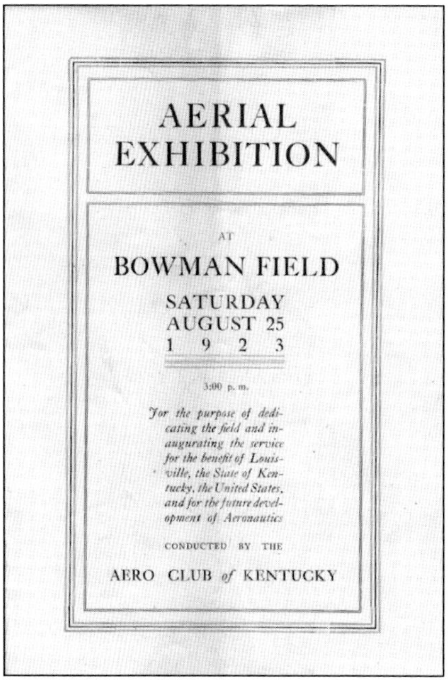

AERIAL
EXHIBITION

AT

BOWMAN FIELD

SATURDAY
AUGUST 25
1 9 2 3

3:00 p. m.

For the purpose of dedicating the field and inaugurating the service for the benefit of Louisville, the State of Kentucky, the United States, and for the future development of Aeronautics

CONDUCTED BY THE

AERO CLUB *of* KENTUCKY

Although Bowman Field was three years old in 1923, it had not been formally dedicated. The dedication, along with an air circus, was held on Saturday, August 25. Widely advertised and with a nice souvenir dedication program, it is estimated that 12,000 attended and saw 50 military airplanes perform quite an aerial extravaganza. At the dedication speech, Abram Bowman was praised for his contributions to local aviation. Speaker Frank B. Russell said, "To the one for whom this field is named is due chief credit for its establishment. Mr. A.H. Bowman saw the possibilities of the future, and largely as the result of his efforts the dedication of this place was made possible . . . the field is dedicated as the Bowman Field. The honor is well deserved." Air Circuses at Bowman Field were so popular that they continued for years. Tickets were eagerly sought. (Both, author's collection.)

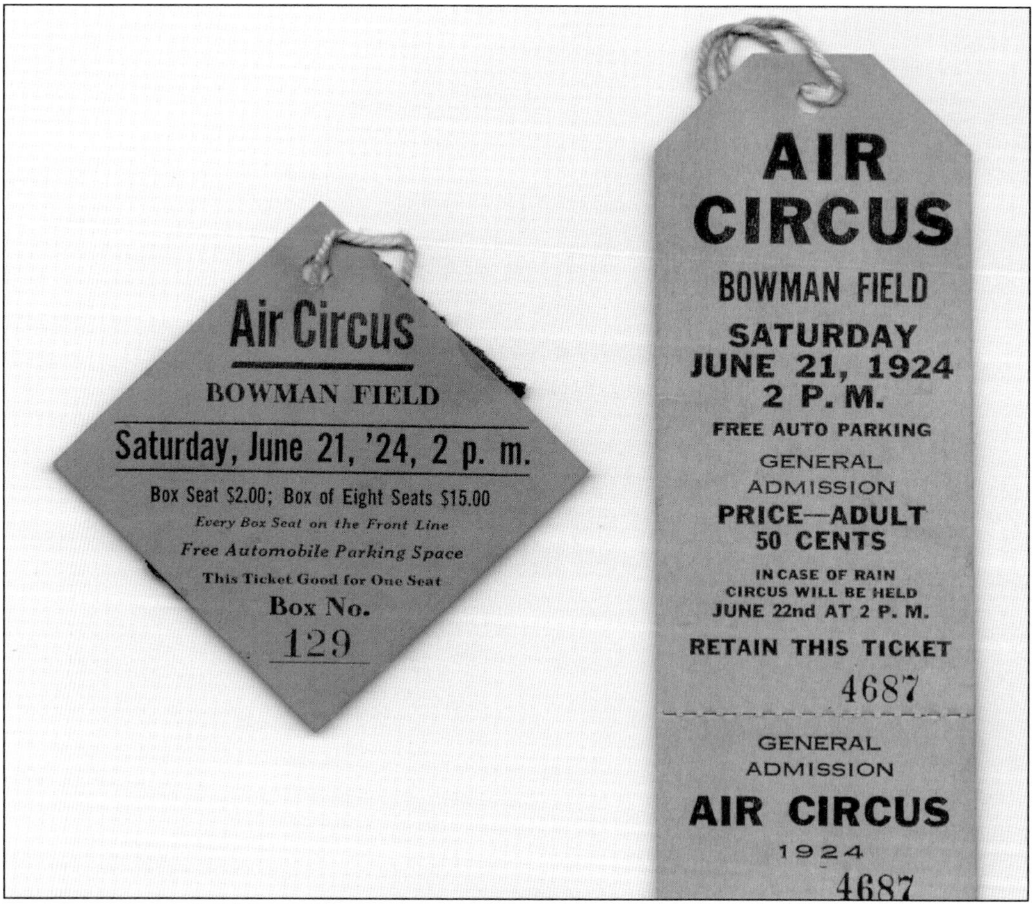

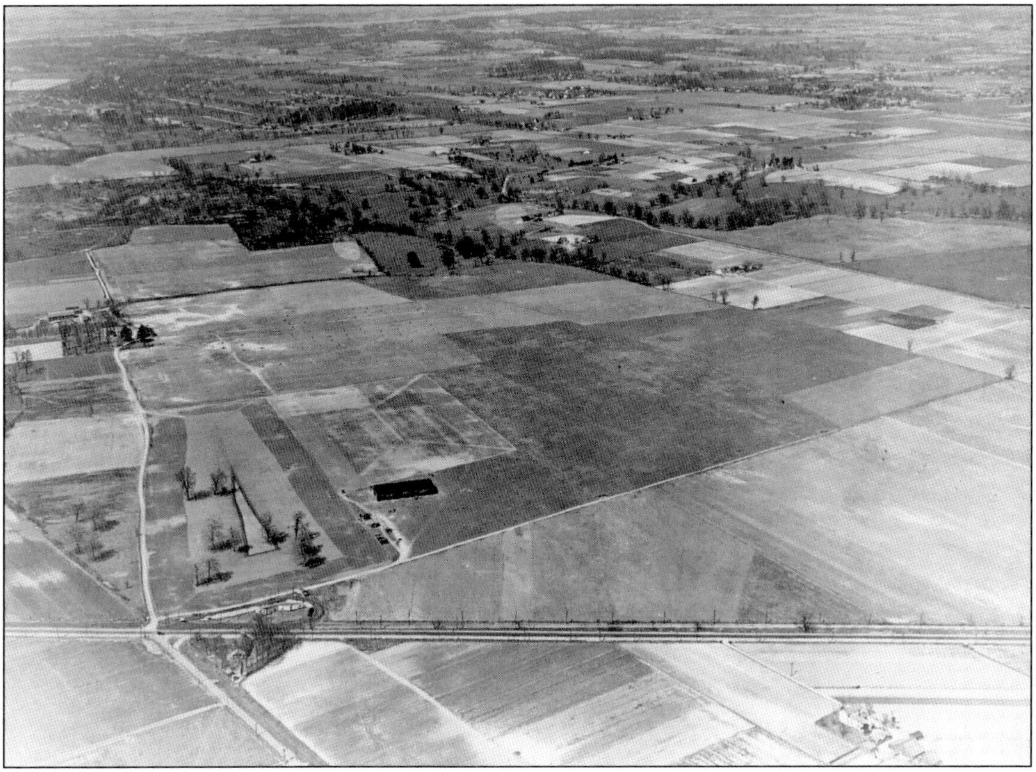

Part of the deal for making Bowman Field an Army Airdrome was moving hangars from Godman Field, Camp Knox, to Bowman Field. Two hangars would be moved. The dismantling and reassembly would be done by members of the Aero Club of Kentucky, with transportation in trucks supplied by Abram Bowman's trucking company. Work started in late 1922 and these early 1923 aerial photographs show the progress. At this time, Bowman Field was 50 acres and the hangars were placed on the west side of the field about where the twin hangars are today. When first erected, the hangars were placed in tandem but were later moved side by side. In addition, three small cottages were also brought from Camp Knox. One burned down, and the other two were eventually moved to where they are today. They are the oldest buildings at Bowman Field. (Above, Edward Peck; below, Chester Lamppin.)

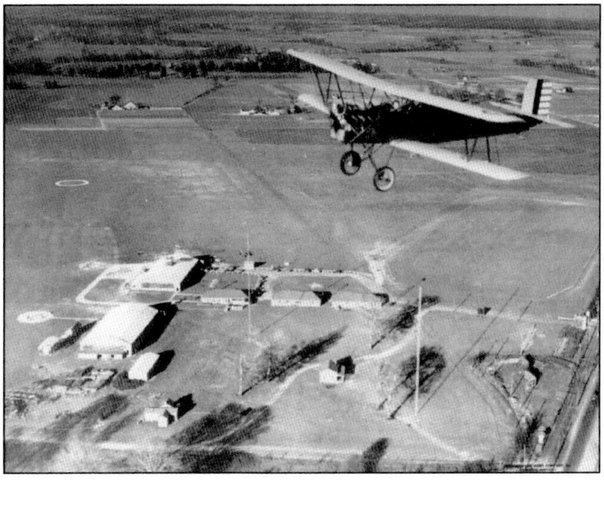

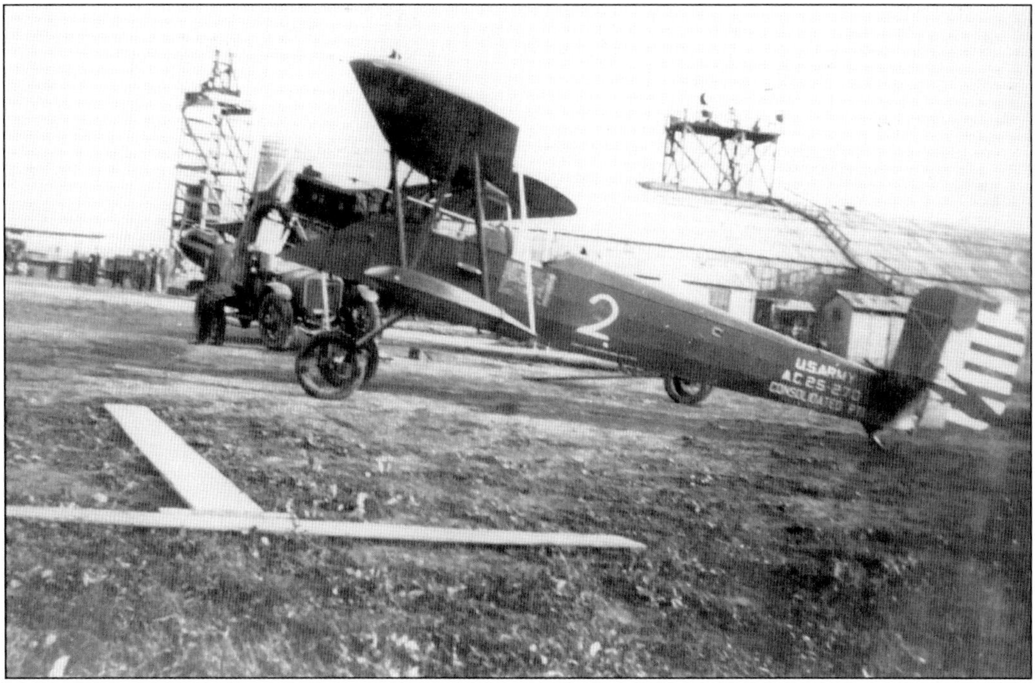

A Consolidated PT training aircraft from the 465th Pursuit Squadron Reserve is shown above on the ground in the late 1920s. Being a reserve outfit, reservist duty was performed on the weekend and the 465th was commanded by a reserve officer, although the Bowman Army Airdrome was commanded by a full-time officer. The first was Lt. Raymond R. Brown. Both of Abram Bowman's partners, Bob Gast and Sidney Park, were members of the 465th. The aerial photograph below shows the tandem arrangement of the metal hangars moved from Camp Knox, along with the two remaining wooden cottages. (Above, Chester Lamppin; below, RAA.)

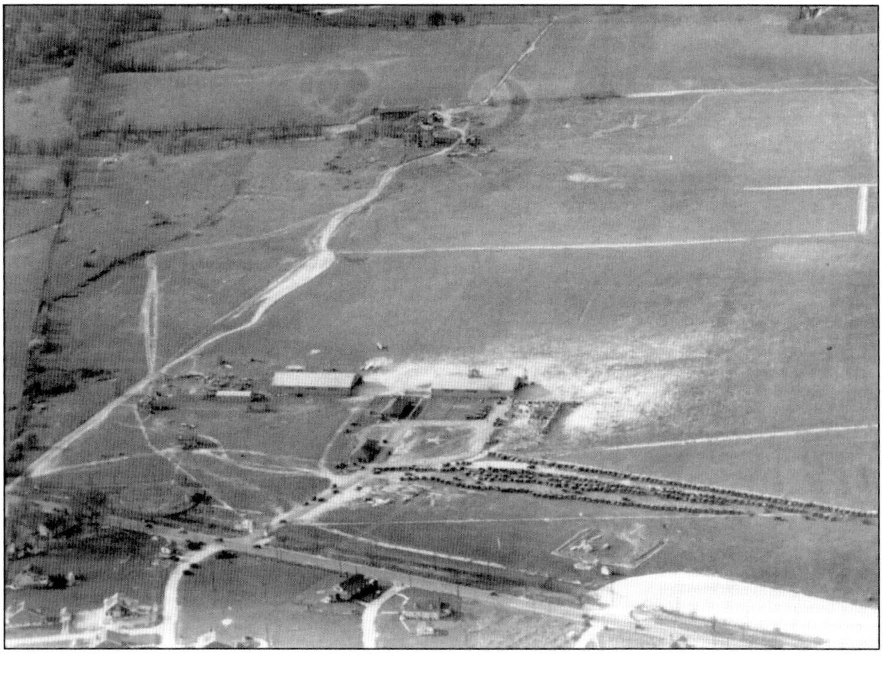

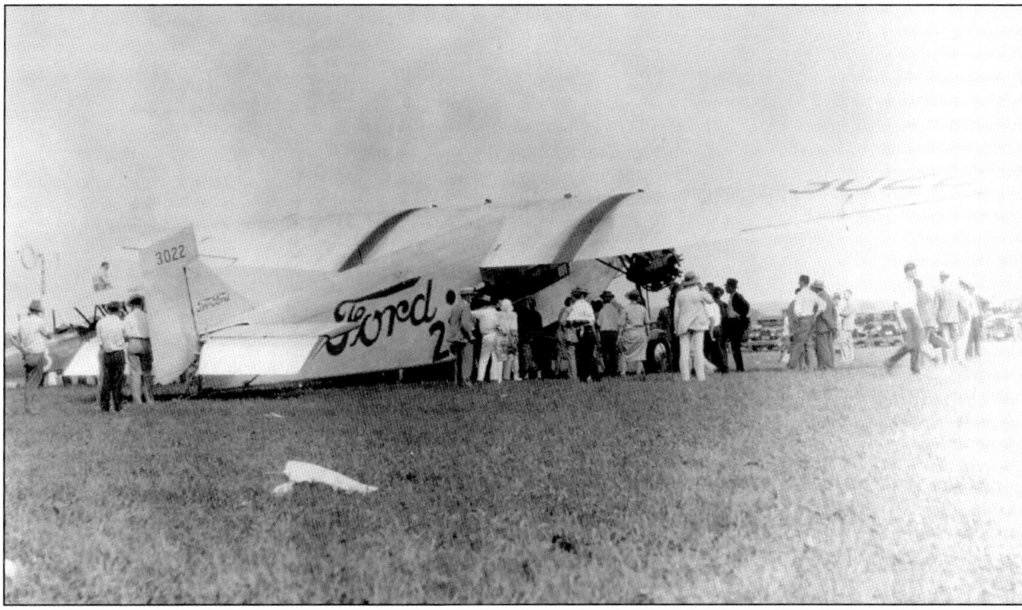

As much as the airplane was a symbol of the future and a new mode of transportation, by the mid-1920s, flying was developing a bad reputation because of poor piloting skills and poor aircraft maintenance. This began to change when Charles Lindbergh proved that an expert pilot and well-maintained airplane could fly safely over long distances. Perhaps the biggest advocate of this was the National Air Tour or Ford Reliability Tour, which started in 1925. This was not a race but a demonstration of pilot skill and aircraft reliability. The tour came to Bowman Field in both 1927 and 1929. Louisvillians were invited to view the tour airplanes when they were at Bowman Field. Here, the exciting new metal airplanes, a Ford Tri-Motor and a single-engine Hamilton All Metal, captivate locals in 1927. (Both, William Russell Beeler.)

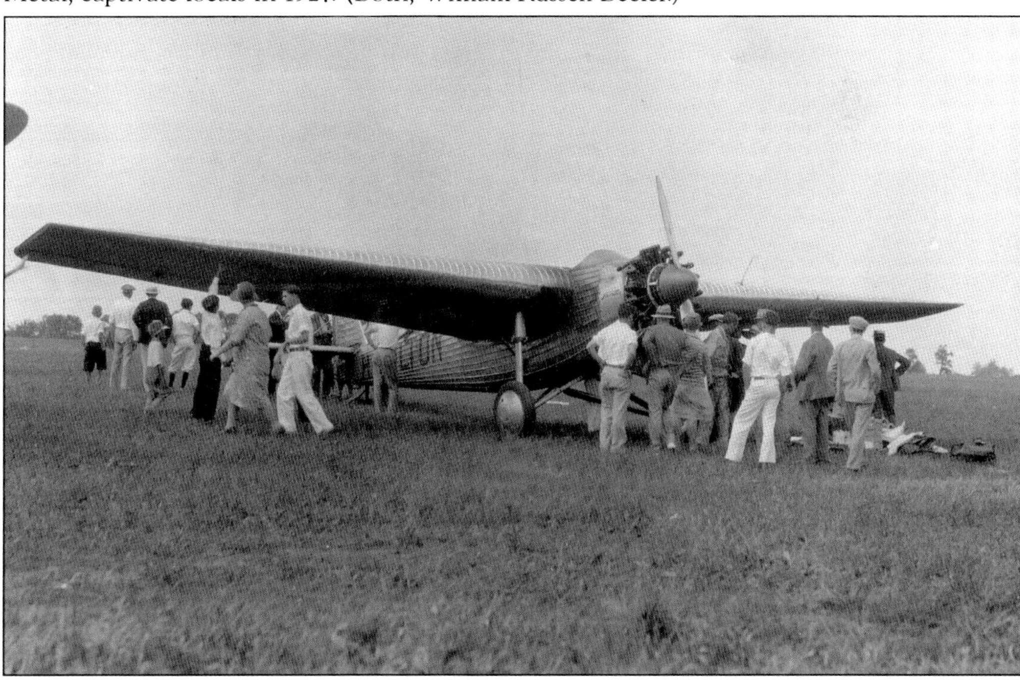

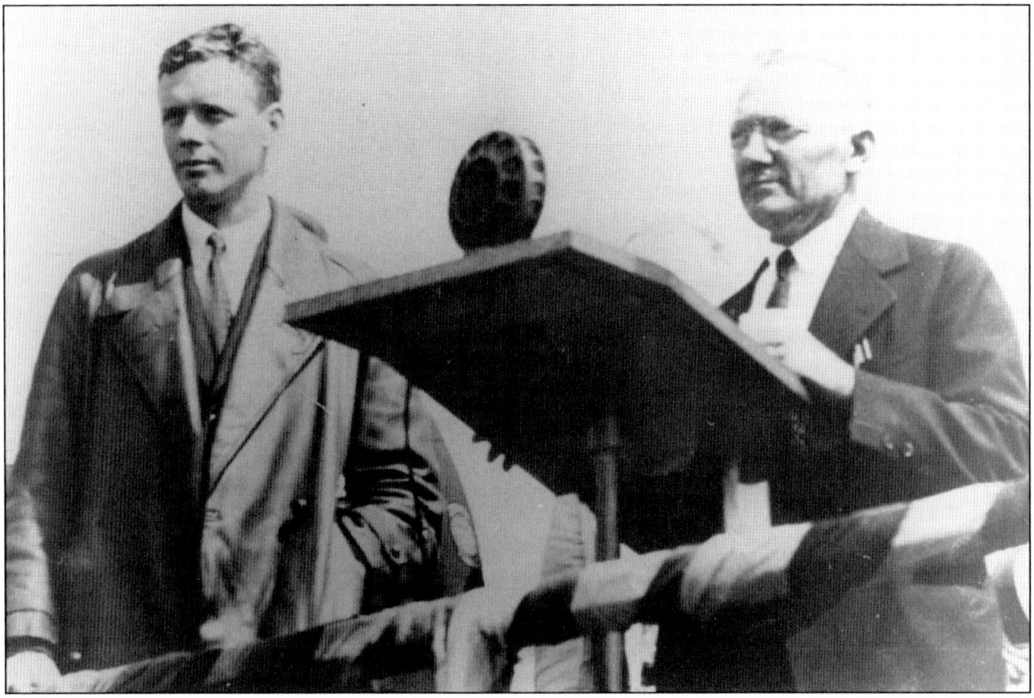

Perhaps the most heart-warming story in all of aviation is the story of Charles Lindbergh's lone flight across the Atlantic Ocean to Paris during May 1927. After the flight, Lindbergh was hailed a hero and dubbed "Lucky Lindy" and the "Lone Eagle." Capitalizing on his fame, the Guggenheim Foundation sponsored Lindbergh to fly a national tour in his *Spirit of St. Louis* airplane starting in July 1927. The intention of the tour was for Lindbergh to promote local aviation in the cities that he visited. Among those welcoming him at Bowman Field was Abram Bowman, when he landed on August 8. A motorcade took him downtown to the Brown Hotel for a dinner and speech that evening. The next morning, Lindbergh flew off to Indianapolis, which was his next stop. (Above, LFPL; below, William Russell Beeler.)

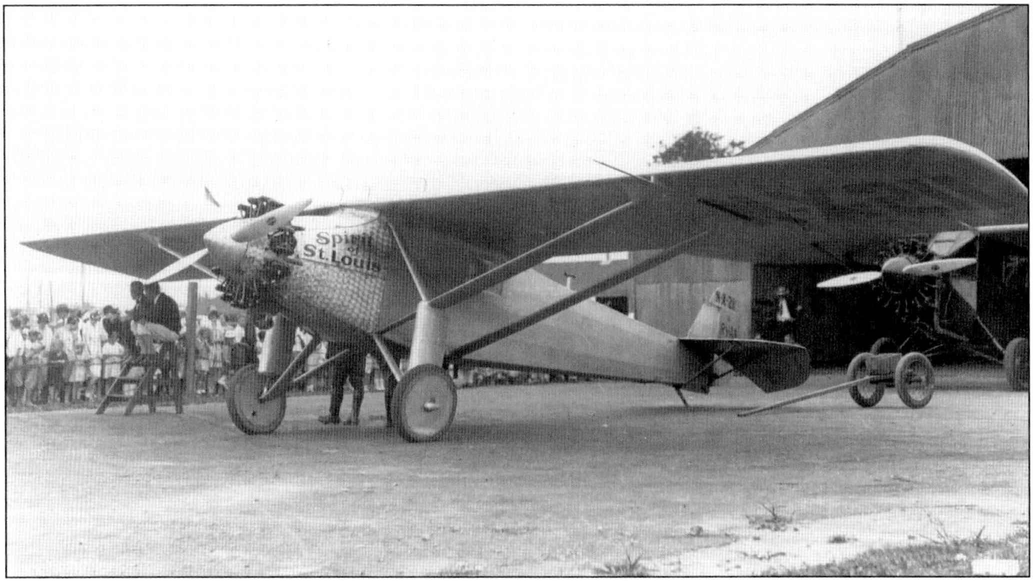

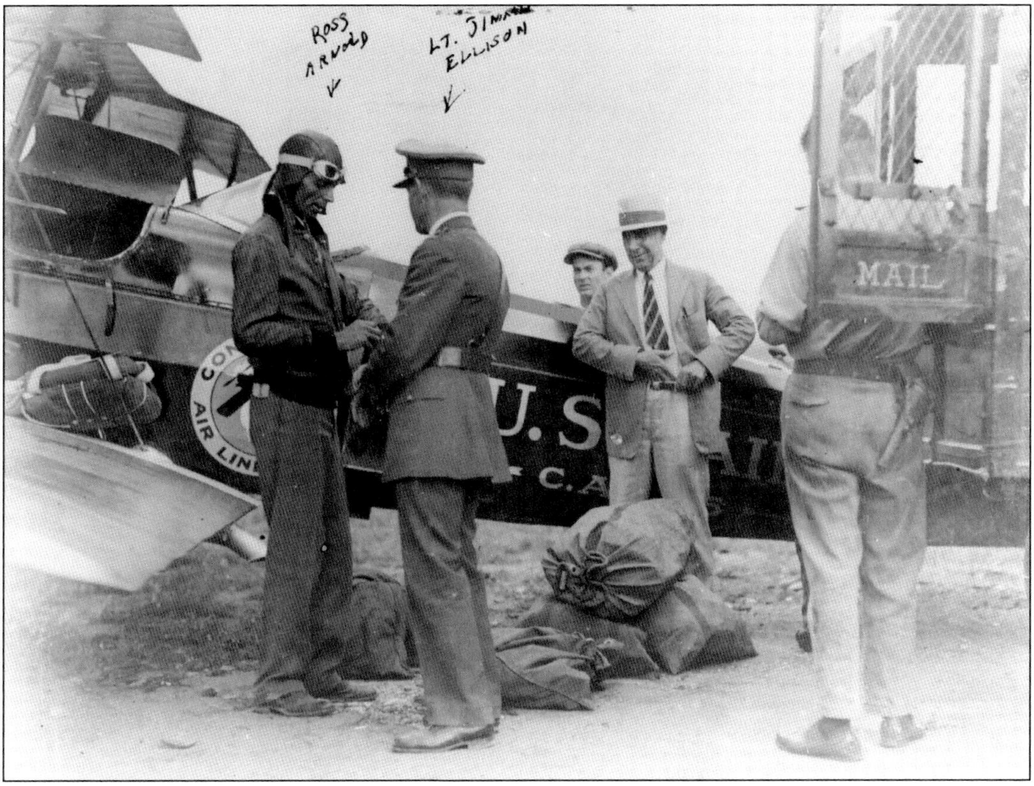

Although Louisville had some earlier air mail service, the first dedicated air mail to the city along the newly established Contract Air Mail (CAM) routes came on August 1, 1928, with CAM Route 16 beginning in Cleveland. On this date, a Continental Air Lines TravelAir biplane piloted by Ross Arnold landed at Bowman Field at 8:00am. Those meeting the arrival included Abram Bowman and Sid Park for the Aero Club; Lt. James A. Ellison, Bowman Field commander; and William E. Robinson, assistant Louisville postmaster. Louisville's hefty outbound air mail for day one totaled 5,200 letters and required two additional airplanes to carry the load. (Chester Lamppin.)

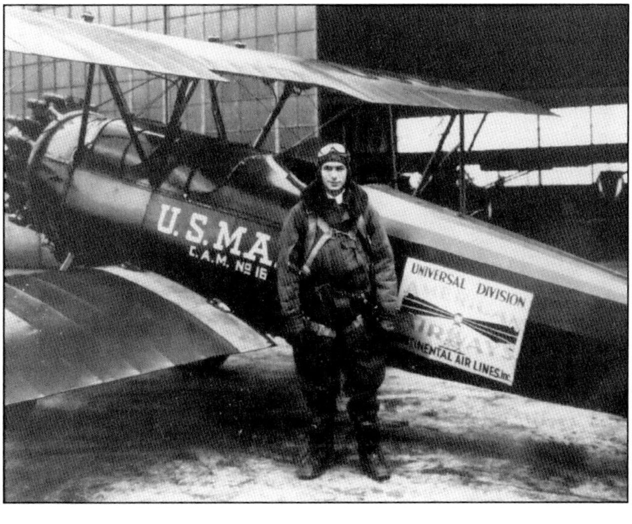

Louisville's first air mail was flown aboard a TravelAir biplane by Continental Air Lines, which, in 1928, was actually a division of a corporation that included American Airways with connections to Lunken Airport in Cincinnati. Other air mail pilots on the early Louisville route were Roy. L. Mitchell (seen here), Marvin O'Dell, and "Indian" Zanpetra. Since the government was subsidizing the Contract Air Mail Routes, airlines actually made more money carrying the mail than passengers. (Chester Lamppin.)

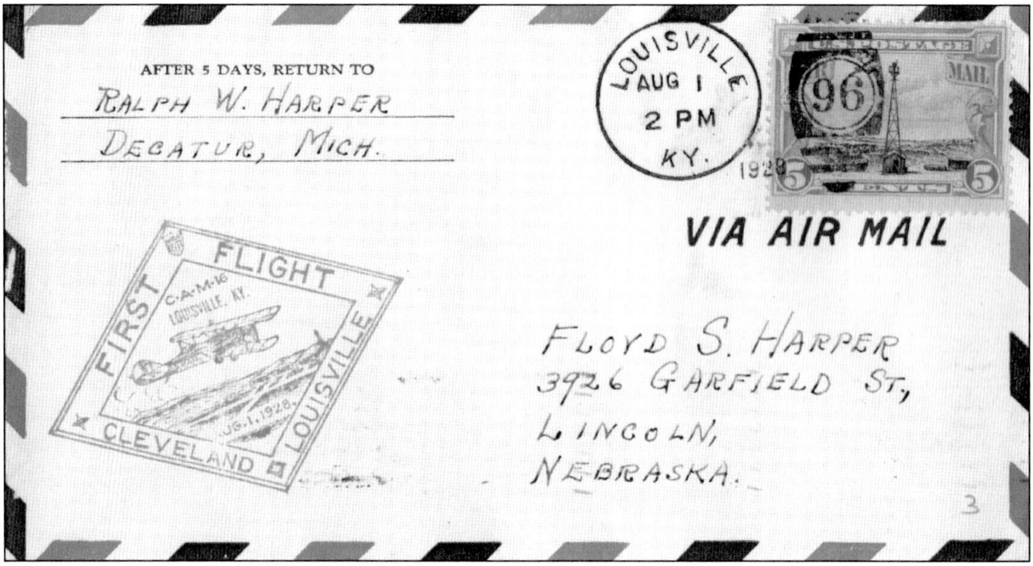

While August 1, 1928, was a big day for Louisville air mail, it was also a big event for United States Air Mail. On this day, a new air mail stamp dedicated to air mail pilots went into effect. The stamp depicted an air mail airplane and one of the many night beacon lights that were mounted on towers and spread out along air mail routes. Perhaps more important for the customer, the price of an air mail stamp was reduced to five cents. In addition to the new CAM Route 16, each piece of mail was stamped with a special cachet to mark the beginning. Other cities along the route were Cleveland, Akron, Columbus, Dayton, and Cincinnati. They all had their own variation of the cachet. (Both, author's collection.)

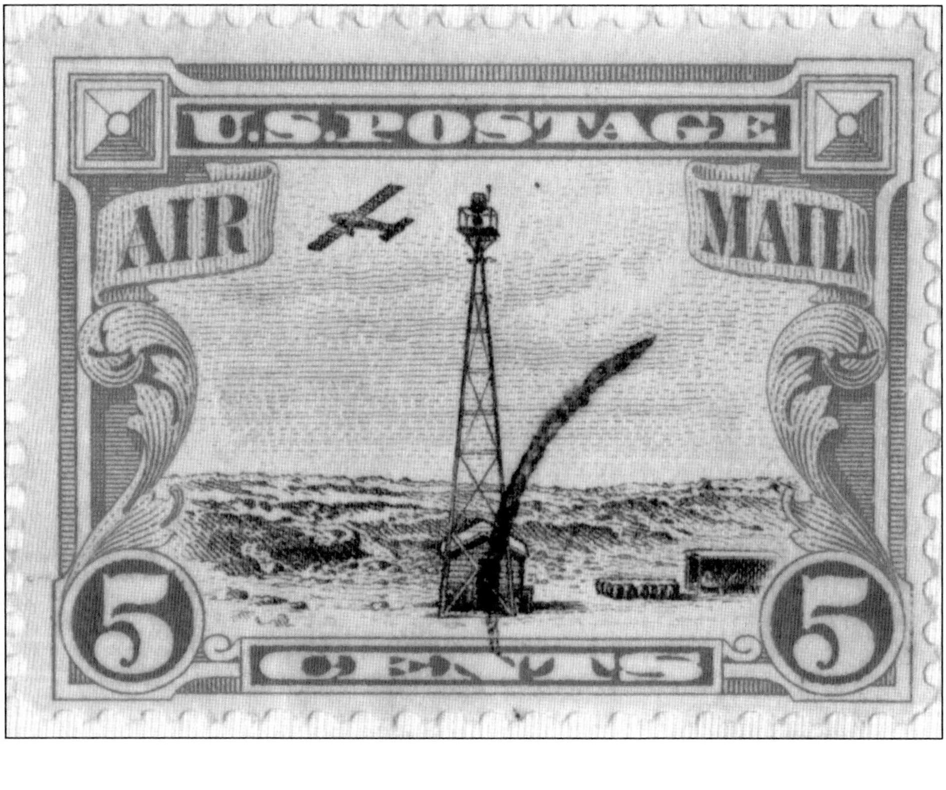

Two

Building Bowman Field

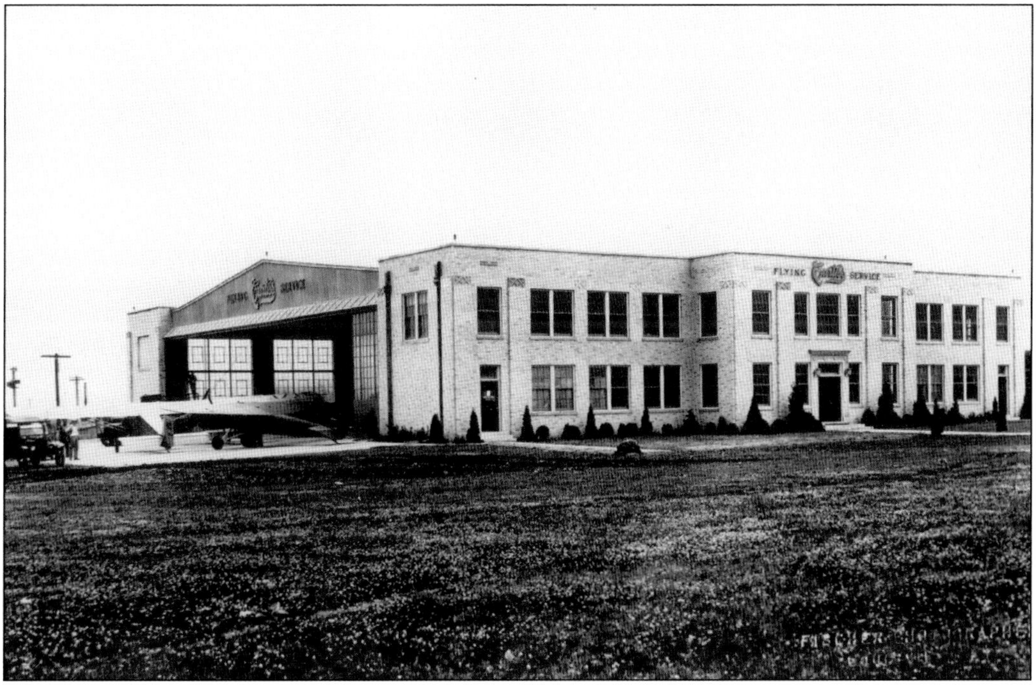

Much like the modern automotive industry uses dealers to sell and service automobiles, Curtiss Flying Service was a pioneer in the aviation industry doing the very same thing with airplanes. Curtiss Flying Service of New York City announced in late 1928 that it was locating to Bowman Field, and it dedicated its building on July 5, 1929. The building had a large hangar on the ground level with classrooms and offices on the second floor. Curtiss Flying Service could offer customers a catalog of airplane brands and types to select for purchase, along with service and hangar space. It could also provide instruction and training all the way to certification for those wishing to become a pilot. Other nearby cities that had Curtiss Flying Service locations included Indianapolis, Memphis, and Columbus, Ohio. (Edward Peck.)

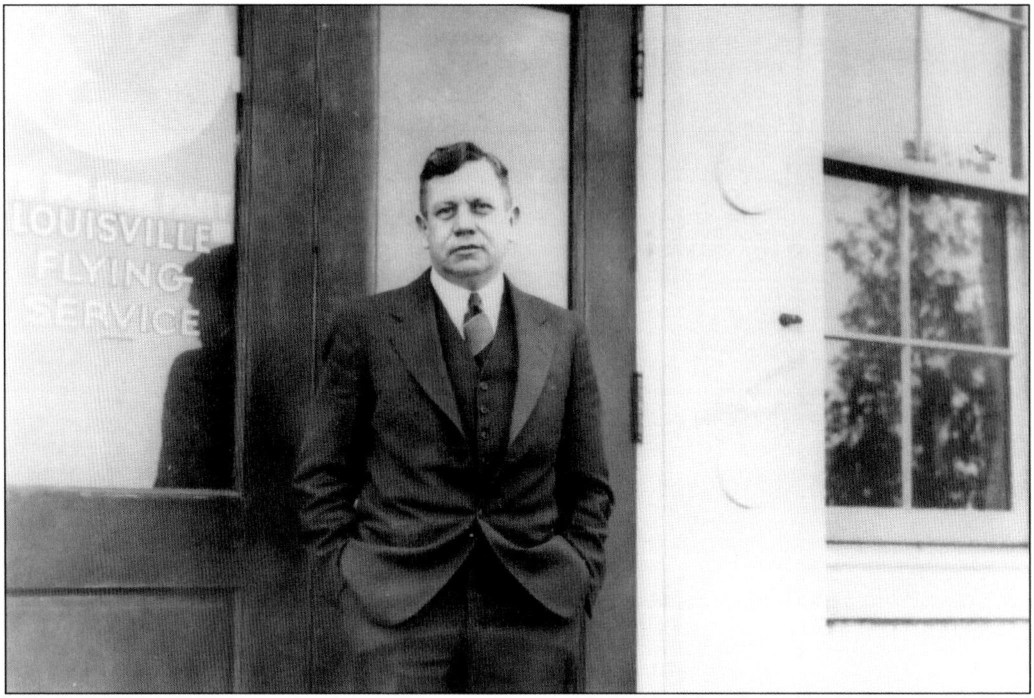

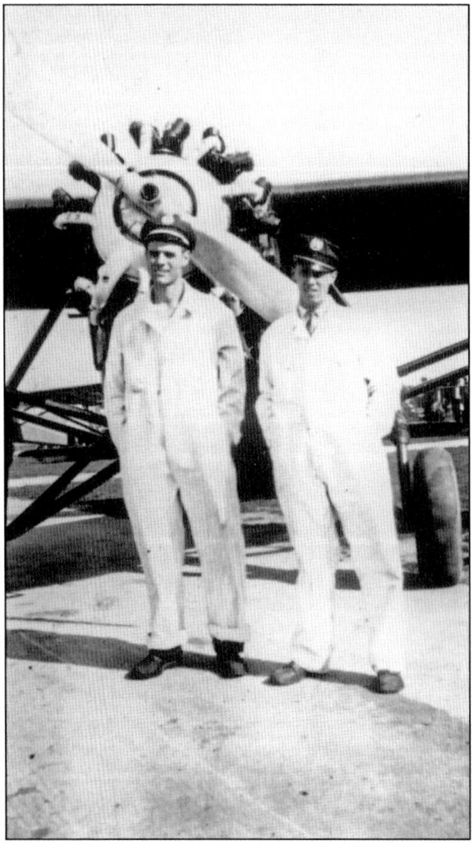

Probably the best thing that happened to Bowman Field before World War II was the arrival of John C. Bennett and Frank Holzaphel. Both were transferred from Nashville by Curtiss Flying Service. During World War I, Bennett served as a pursuit pilot, and after the war in Nashville, he organized the 136th Aero Squadron of the Tennessee National Guard, obtaining the rank of major. He preferred to be called "Major," or better yet, "Maj." In 1932, as manager of Bowman Field's Curtiss Flying Service, he reorganized as Louisville Flying Service (LFS) when Curtiss departed as a Depression-era economy move. Aircraft mechanic Frank Holzaphel (left), vice president and service manager at LFS, was well known for his mechanical skills. Both took an active role in Louisville aviation and maintained a quality flying service during the Great Depression. (Both, William Russell Beeler.)

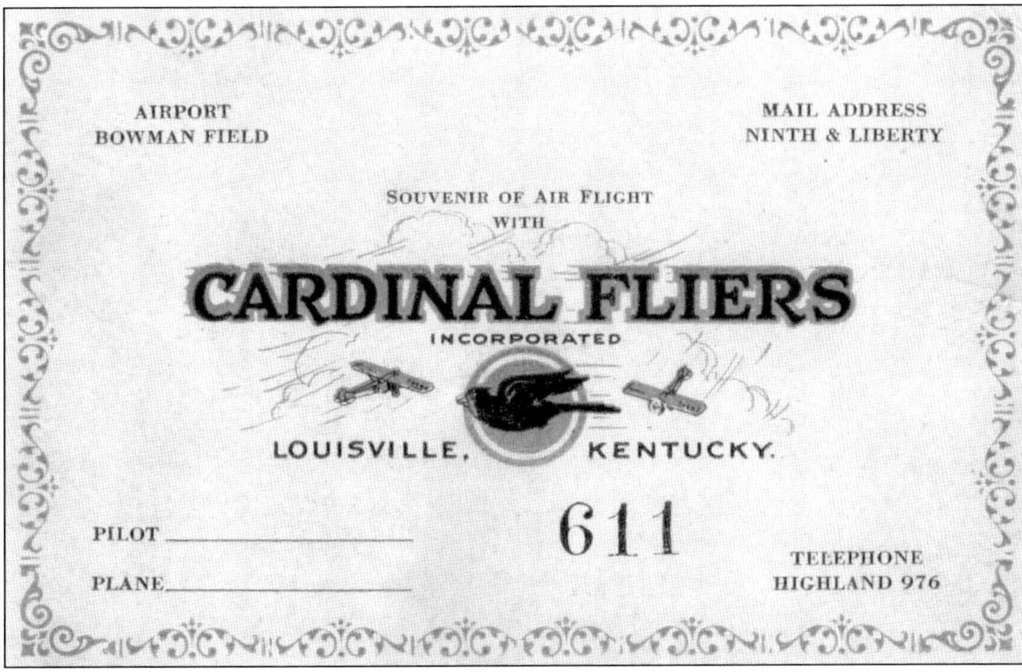
AIRPORT
BOWMAN FIELD

MAIL ADDRESS
NINTH & LIBERTY

SOUVENIR OF AIR FLIGHT
WITH

CARDINAL FLIERS
INCORPORATED

LOUISVILLE, KENTUCKY.

611

PILOT ______________________

PLANE ______________________

TELEPHONE
HIGHLAND 976

Cardinal Fliers, which was incorporated on March 15, 1928, was Bowman Field's first true air service that offered amenities comparable to modern day counterparts. It was operated by Lee L. Miles and had Bob Gast as general manager and chief pilot. Cardinal Fliers was preceded by a smaller Yellow Air Taxi Service two years earlier run by the same people. It offered student training, sightseeing flights, aerial photography, and what might be called "on demand" air service to Lexington, Kentucky. It used and sold WACO airplanes and had a large Fairchild aircraft to carry passengers on charter flights. In late 1928, it was announced that Curtiss Flying Service would absorb Cardinal Fliers to make room for its new operation the next year. (Above, author's collection; below, RAA.)

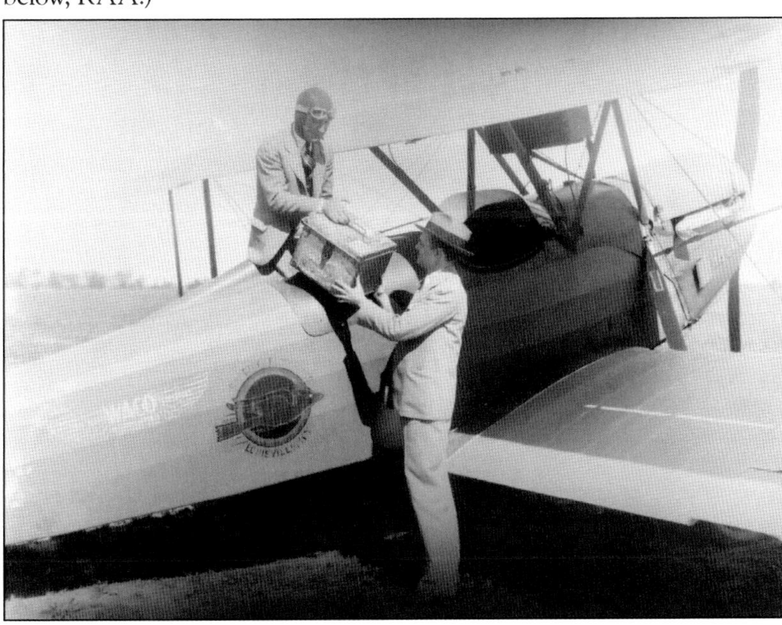

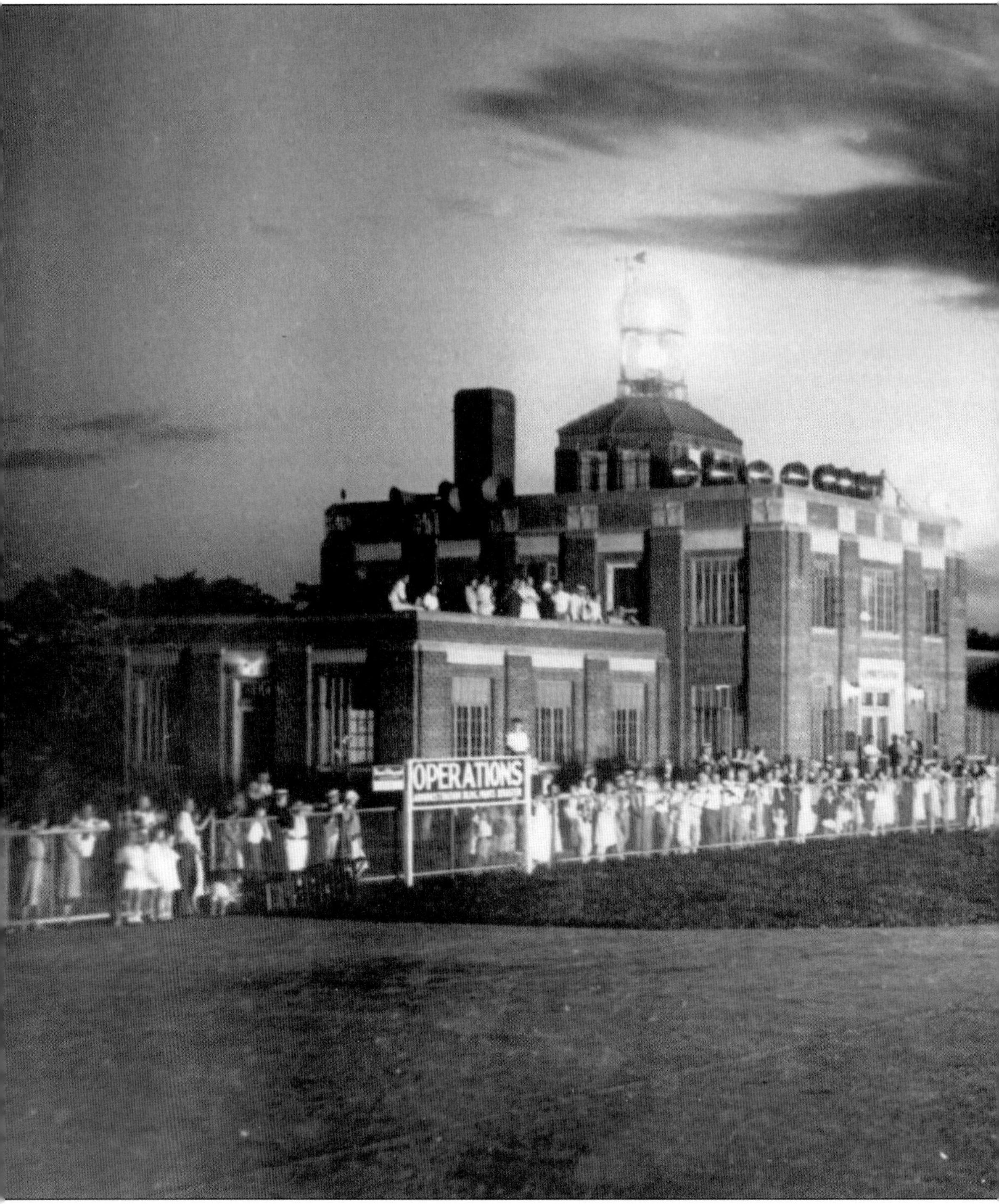

Dusk at Bowman Field during 1932 shows an American Airways Ford Tri-Motor arriving or preparing for takeoff. American Airways was the first major passenger-carrying airline at Bowman Field, starting service in 1931. The Ford Tri-Motor has been called the first modern American airliner. The thrill of seeing airplanes has brought out a group of spectators to watch the flight.

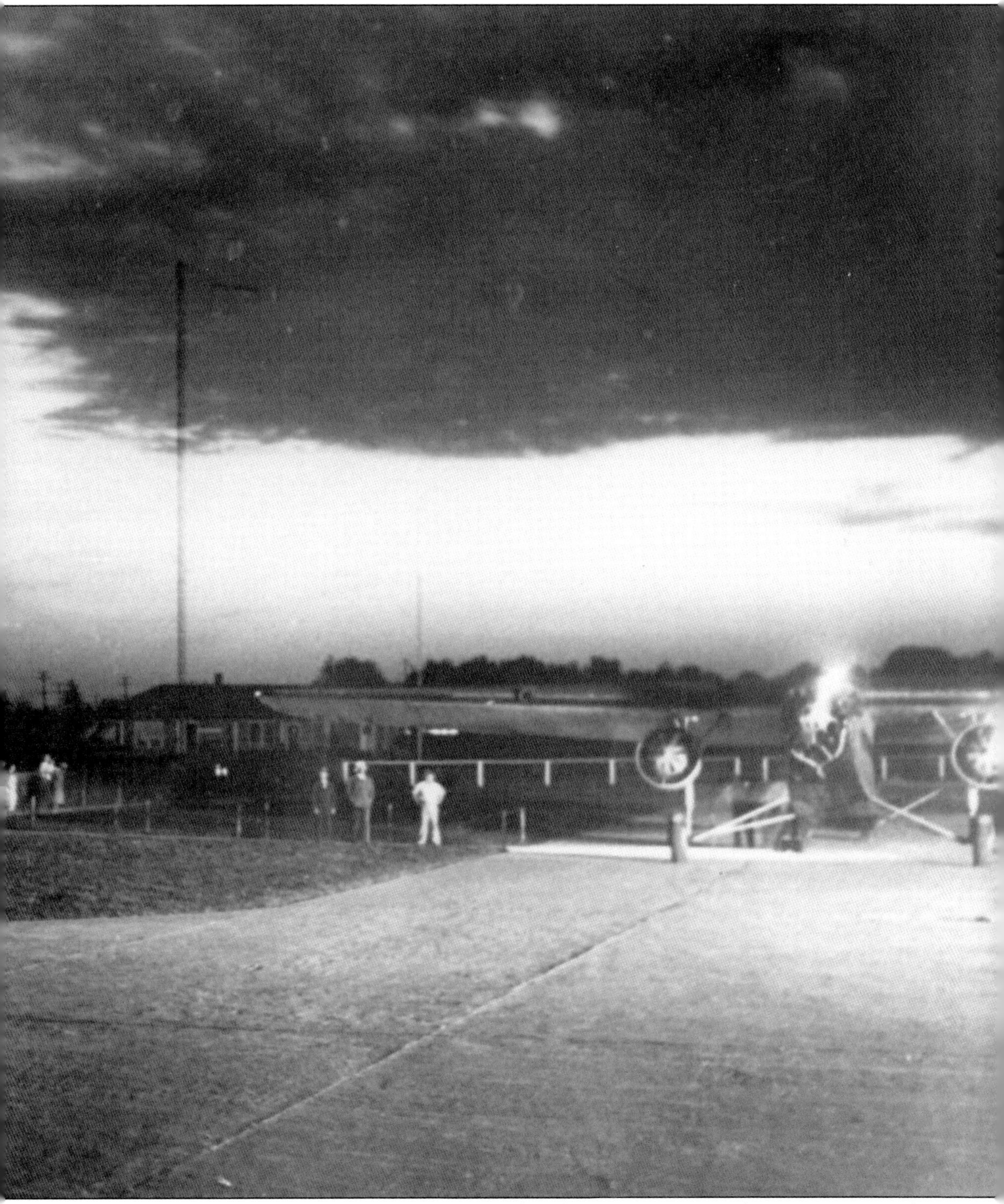

During a time when most long-distance travel was done by passenger train, there was a sense of awe and wonder for anyone traveling by airplane. Also of note, Bowman Field's tall radio antenna is seen in the background. (LFPL.)

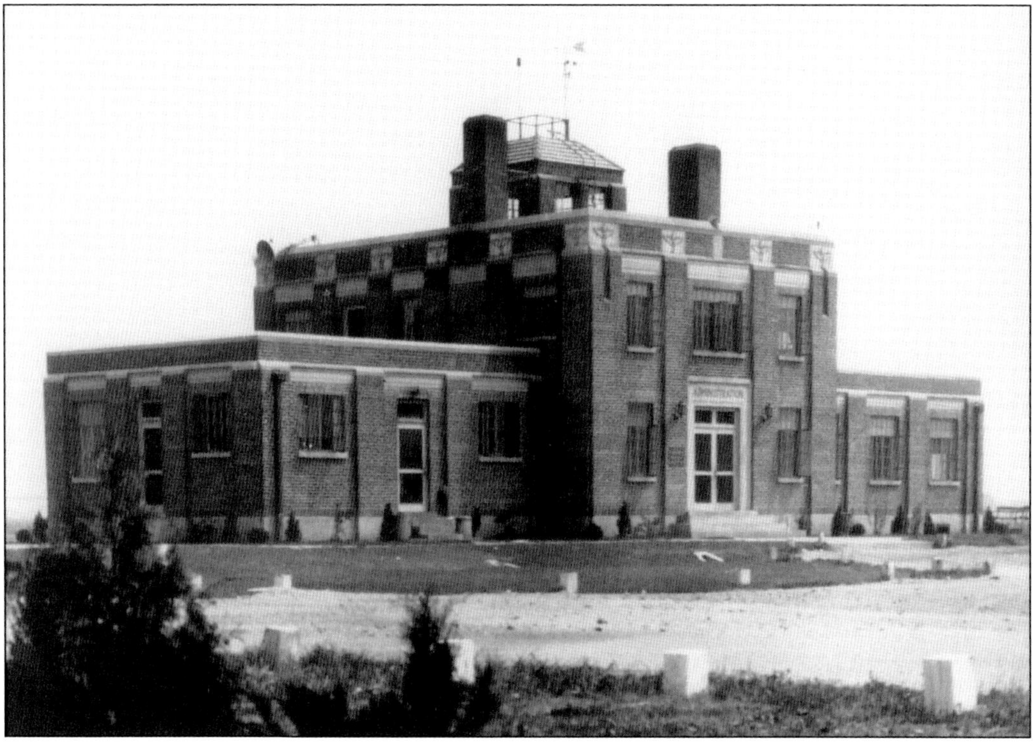

After the creation of the airline industry in the United States during the 1920s, it was only a matter of time before it came to Bowman Field. Construction started in 1929 on a beautiful $25,000 administration building. Designed by noted architect W.S. Arrasmith, it had a large center section with two smaller wings on each side. The building was dedicated on October 2 and featured a restaurant and offices for the air board as well as the weather bureau. This redbrick and concrete building with a large parking lot was positioned along Taylorsville Road next to Curtiss Flying Service on the airport's southwest side. (Both, Edward Peck.)

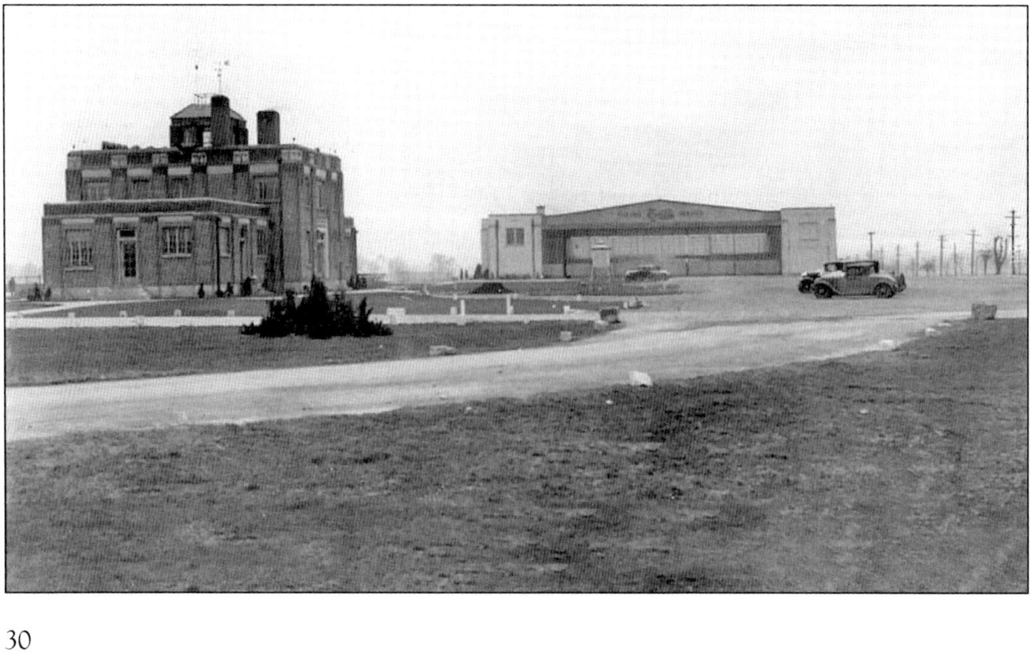

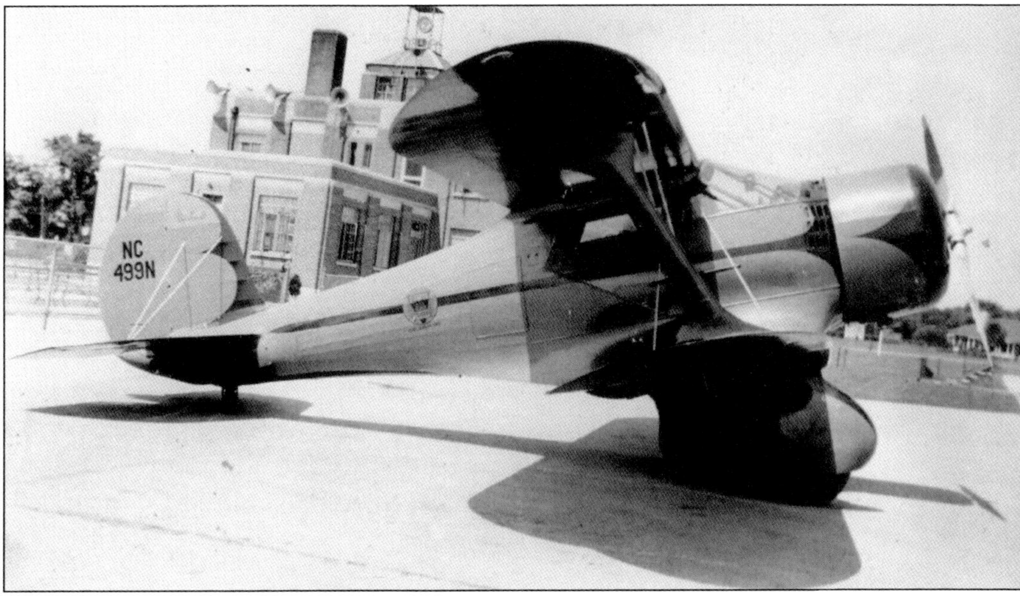

Airports are built to attract airplanes, and over the years, Bowman Field has had its share of famous and interesting ones. The name Walter Beech resonates through the aircraft industry for his company's many famous designs including the Staggerwing. Its fast speed and stunning looks made it a success, with claims as the most beautiful biplane of all time. The Staggerwing shown above first flew in 1932 and is seen at Bowman Field shortly after. It was the first one made and had fixed landing gear, while production Staggerwings had retractable landing gear. This airplane is currently on display at the Beechcraft Heritage Museum in Tullahoma, Tennessee (below). (Above, Edward Peck; below, author's collection.)

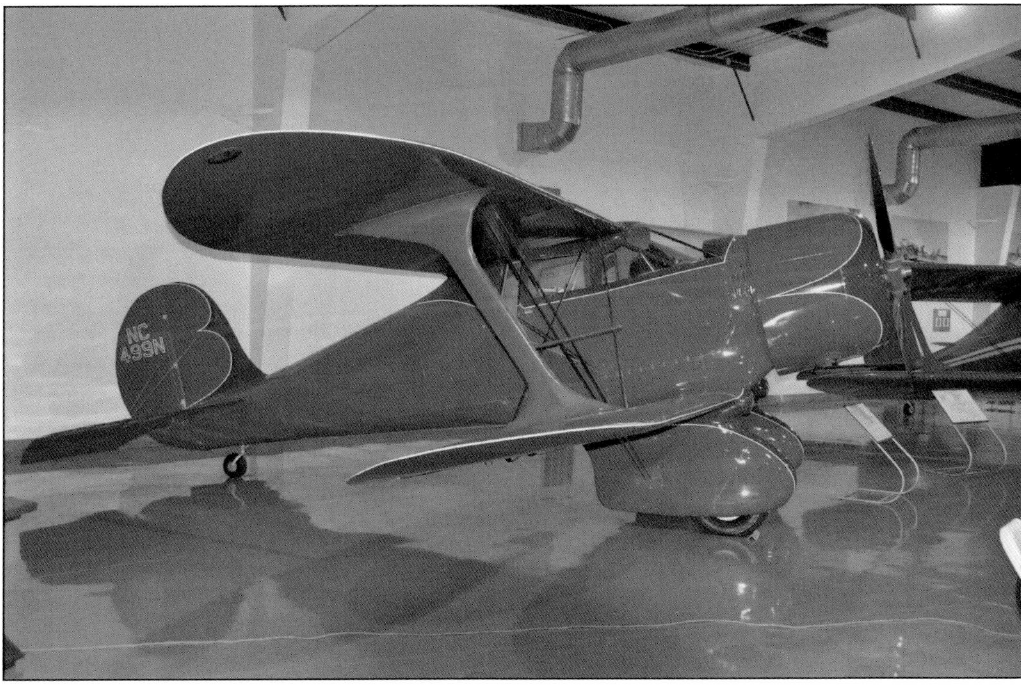

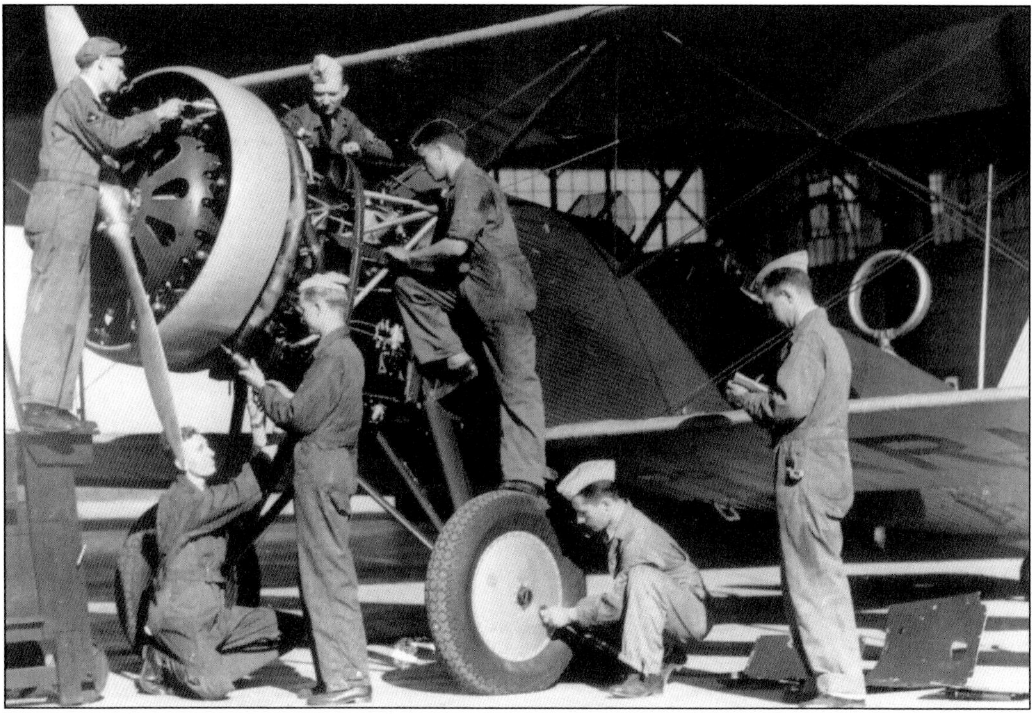

During the late 1920s, the mission of the Bowman Field reserve squadron changed from pursuit to observation. It was no longer the 465th Pursuit Squadron Reserve but now the 325th Observation Squadron Reserve. In addition, the Army Air Service changed its name to Army Air Corps. Pictured above are enlisted mechanics from the 325th: Paul Echsner (with notebook), Henry Julliard (checking tires), Francis Blanford (standing with grease gun), Albert Washer (kneeling), Samuel D. Martin (on stand), Chester Lamppin (above firewall), and unidentified (on wheel). Identified men standing in front of the airplane below include Al Near (standing center), Al Woody (standing right), and Robert McDaniel (kneeling right). (Above, Chester Lamppin; below, Edward Peck.)

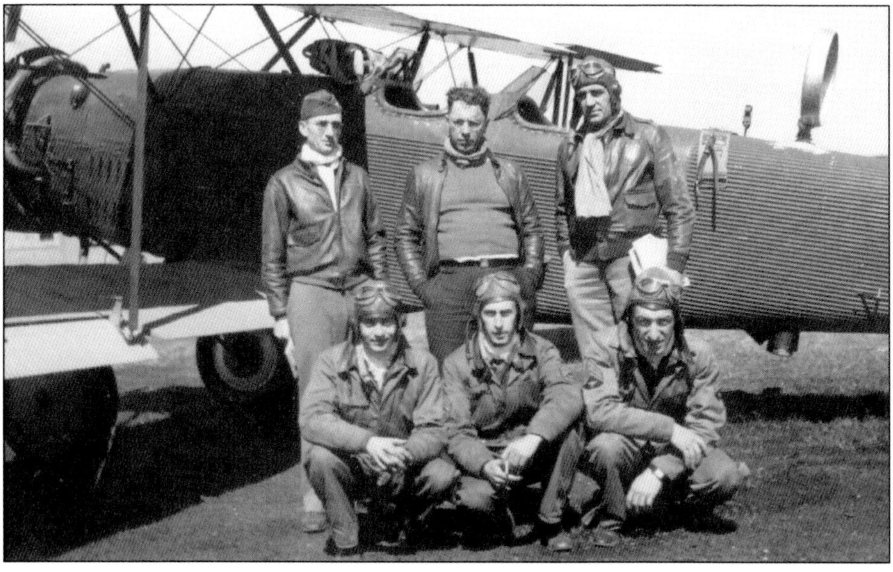

Lt. Al Near poses in front of a 325th Observation Squadron Reserve airplane. Near is a good example of an early Bowman Field leader who wore two different hats, one for the military side of the airport and one for the civilian side. In addition to his part-time role as a pilot in the reserve squadron, he was also employed full time by the Louisville and Jefferson County Air Board as superintendent of Bowman Field. His able leadership helped transition Bowman Field to a military airfield in 1940 and back to a civilian airport after World War II. (Edward Peck.)

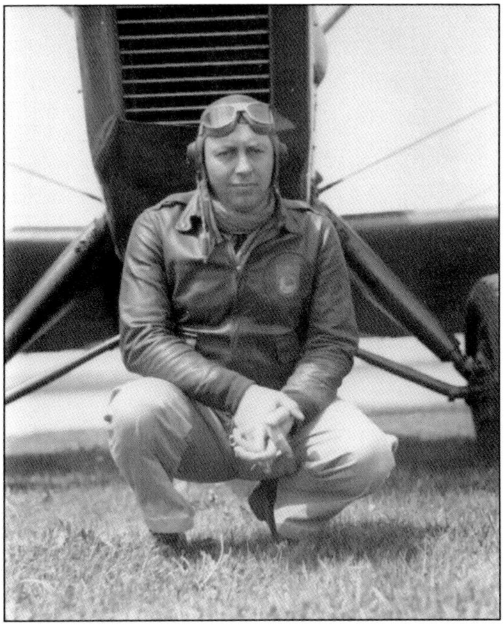

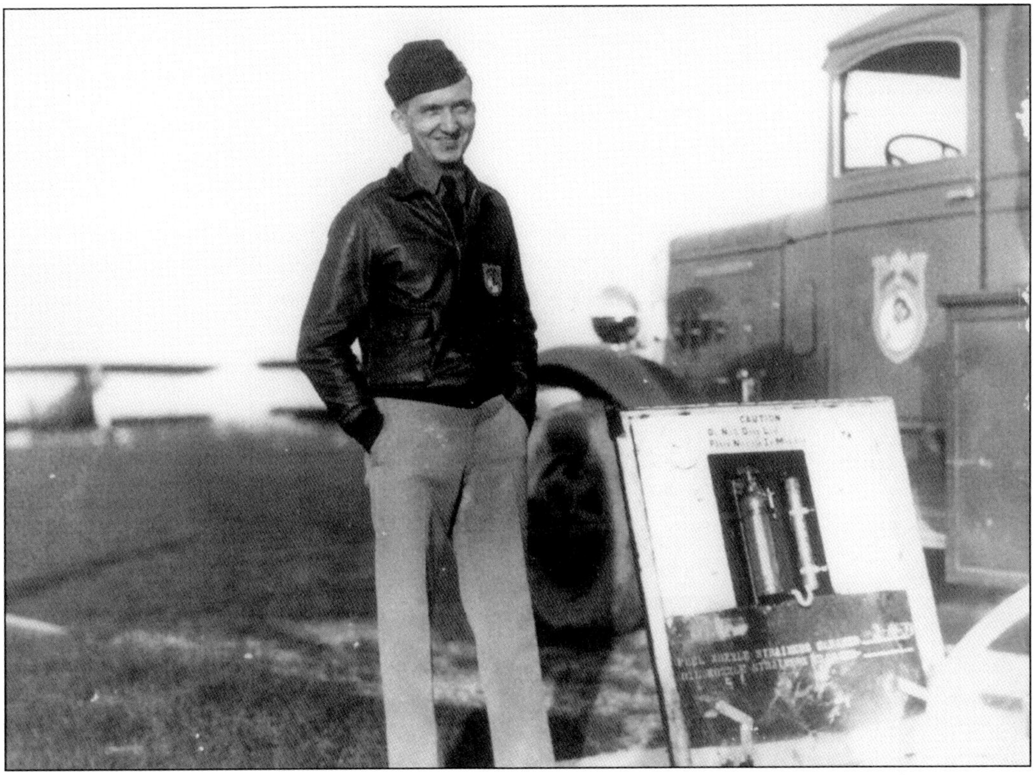

Sgt. Chester Lamppin had quite an extensive career with the 325th Observation Squadron. Signing on during the 1920s as an aircraft mechanic, he eventually became one of the few full time reservists putting aside his tools for a typewriter to do administrative duties. He truly loved being a member of the 325th and his excellent memory of people and events added much to understanding the history of the unit. Several of his photographs appear in this book. (Chester Lamppin.)

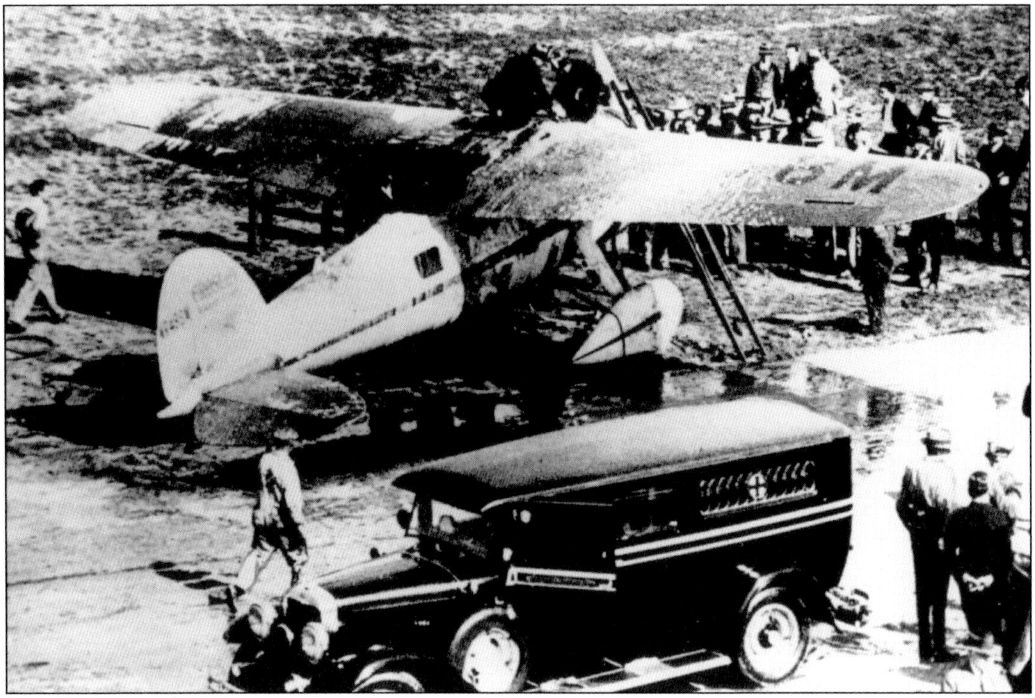

During the 1930s, women were in the forefront of aviation progress by setting new speed and distance records. One who made headlines was Ruth Nichols. Beginning in 1930, she was sponsored by Powell Crosley in Cincinnati and used his Lockheed Vega airplane for many of her flights. On October 24, 1931, Nichols departed Oakland, California, for New York to attempt a distance record for women. She landed in Louisville on October 25 after setting a record of 1,977.6 miles. Preparing to take off from Bowman Field the next morning, her Vega caught fire while on the ground, ending the rest of her planned flight. Nichols was unhurt, and her airplane was rebuilt. (Above, Edward Peck; below, author's collection.)

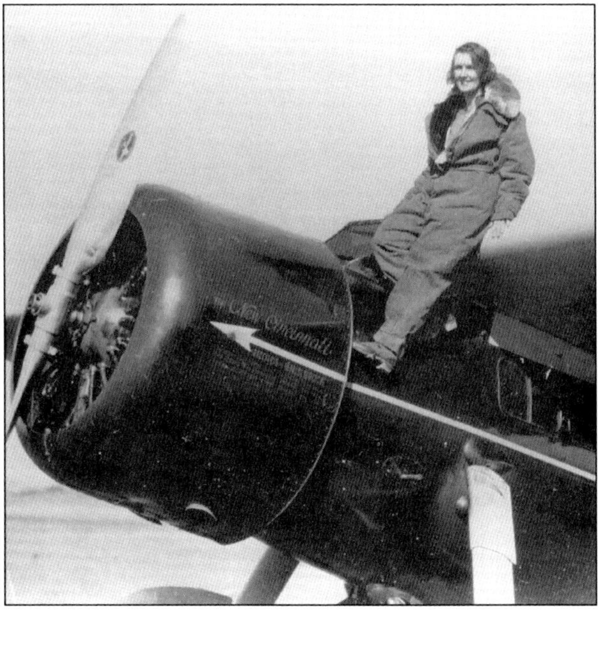

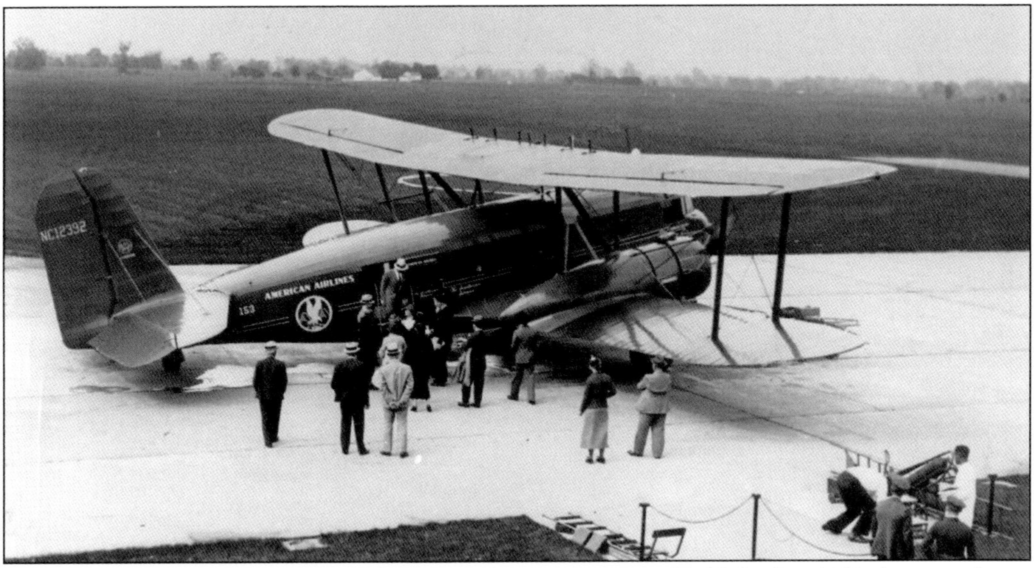

The Ford Tri-Motor was not the only early airliner at Bowman Field. In this 1932 photograph, an American Airways Curtiss Condor awaits its next flight. Although the Condor was a biplane, some had retractable landing gear and could be configured as a sleeper for night flights. Note that passengers and crew had to walk to and from the airplane regardless of the weather. Furthermore, no paved runways are visible; the airplane took off and landed in the grass. (LFPL.)

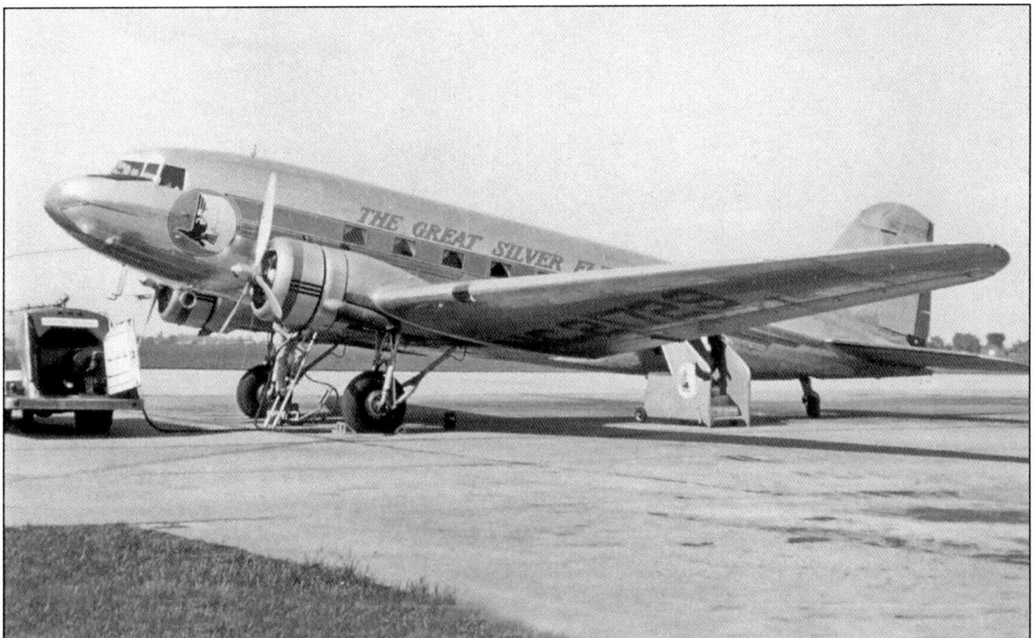

An Eastern Air Lines Douglas DC-3 is shown at Bowman Field during the late 1930s. In time, all of the major airlines at Bowman Field would fly the DC-3. It is the airplane that revolutionized the airline industry because of its safety and reliability. Equally important, it could carry 21 passengers and, for the first time, allow airlines to make a profit flying passengers without relying on air mail subsidies. After World War II, corporate DC-3s were based at the field, and it is not unusual to still see one from time to time. (Edward Peck.)

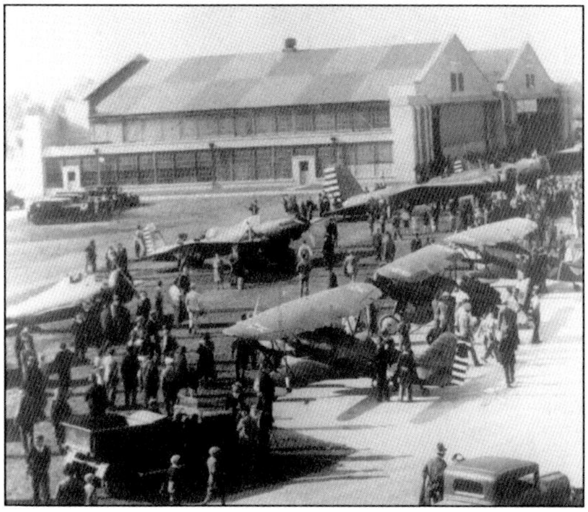

Known as the twin hangars, these two attached hangars were dedicated on October 22, 1932. Positioned on the west side of the airport, they were built for the 325th Observation Squadron Reserve and replaced the two metal hangars erected in the early 1920s. With public participation, the dedication included a large military aircraft display. Built with steel, concrete, and brick, they were released after World War II for civilian use. The first to use the hangars was Hoop Davies with his Davies Flying Service, which in 1950 gave way to Louisville Flying Service when it moved from the old Curtiss Flying Service hangar. (Edward Peck.)

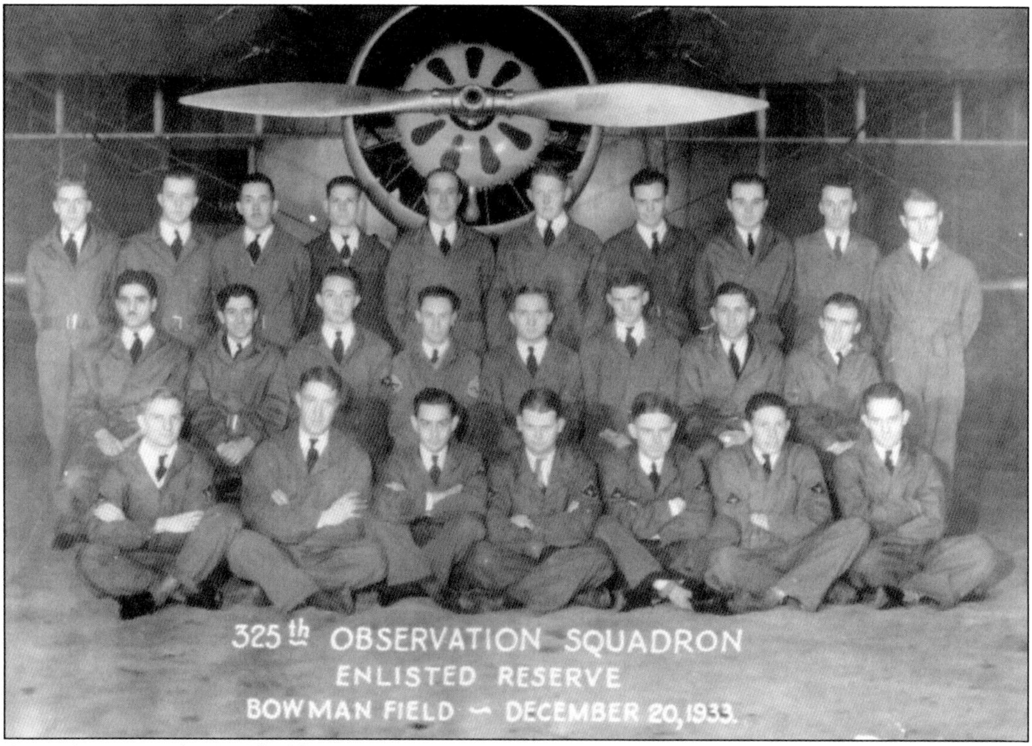

Over the years, the 325th Observation Squadron Reserve held at least two annual get-togethers, a summer picnic, and Christmas party. This photograph shows enlisted members at the hangar before one of the events. While the Christmas party was held at a local hotel, the picnic was held at the airfield or a patron's farm. The Bowman squadron, like most reserve outfits between the wars, was not large in number. Aircraft were simpler to maintain in those days and maintenance personnel could do a variety of jobs. (Chester Lamppin.)

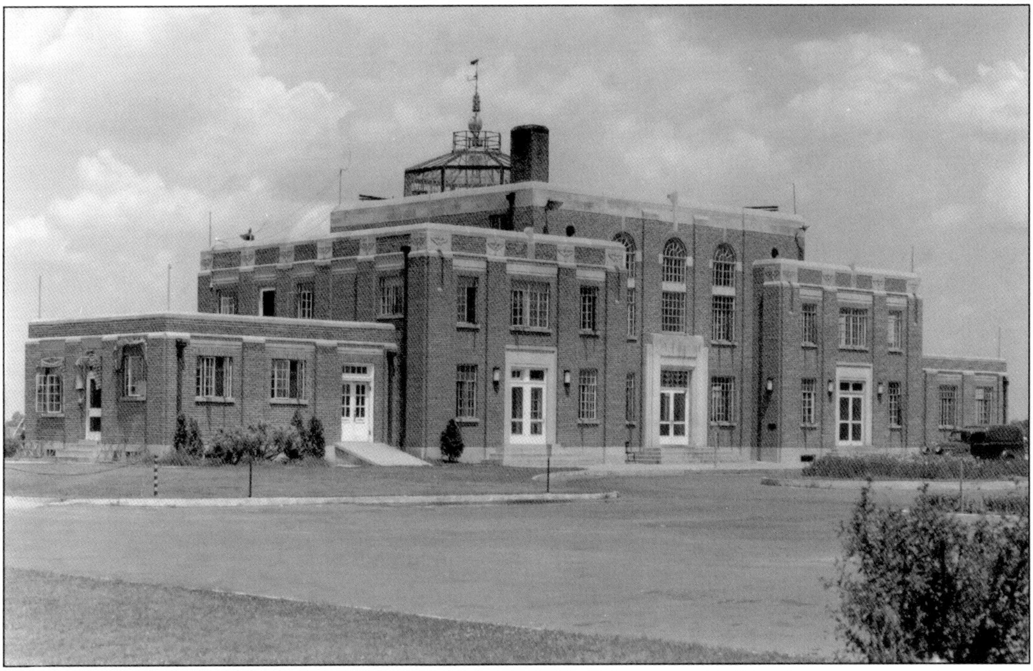

During 1936, it was decided to enlarge the size of the Bowman Field administration building. The work retained portions of the existing building for the west side and added a large center section plus duplicating the west portion for the new east side. The center area featured a large open lobby with airline ticket windows and passenger seating. Offices were on the second and third floors and a restaurant returned to the east side. A glassed-in air traffic control tower was on top of the building. It was rededicated in June 1937. This beautiful building stands today and is among the finest examples of early airport architecture in the United States. It is well worth a visit. (Both, Edward Peck.)

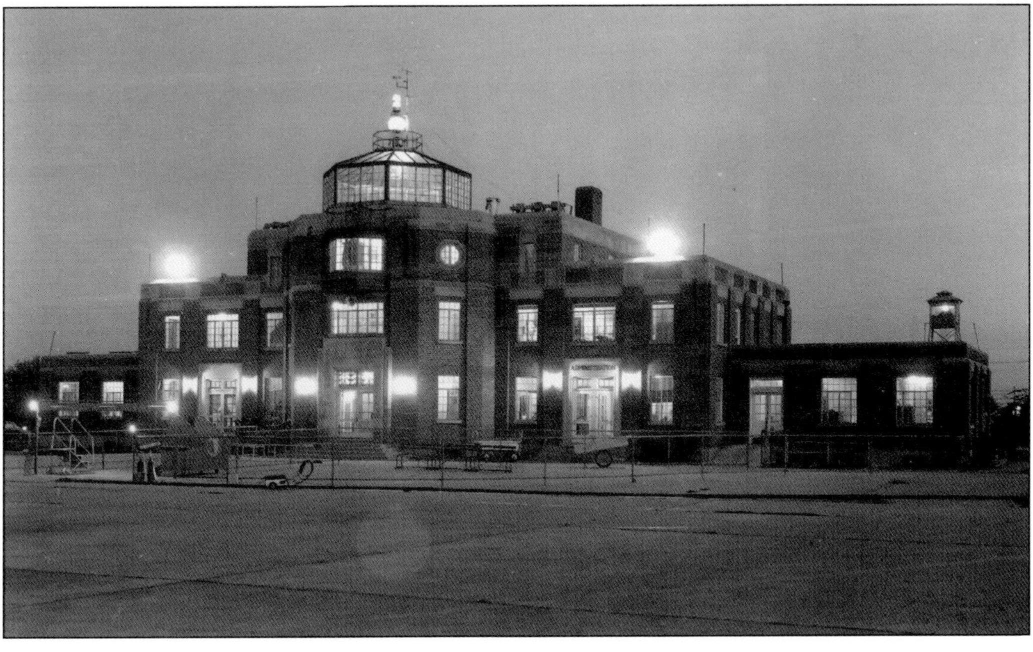

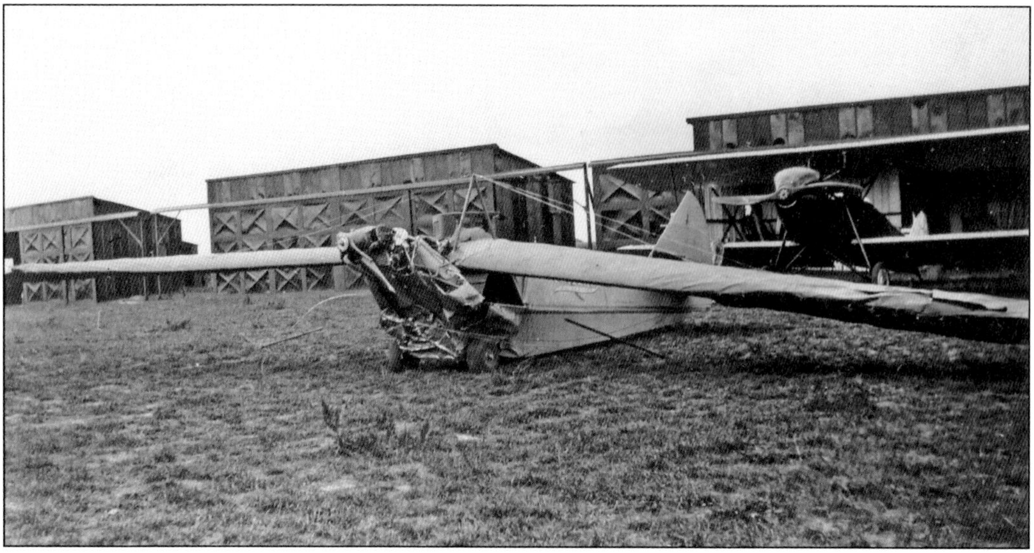

Crescent Air Service started operation at Bowman Field during June 1929 with John Lee Dunlap as president, Pat Moran as vice president, and Lap Ewing as secretary/treasurer. It was in business to buy and sell airplanes, provide charter and sightseeing services, and do aerial photography. On Sunday, October 5, 1930, during an impromptu air show at Bowman Field, Crescent's Aeronca C-1 airplane piloted by Pat Moran crashed after being caught by another airplane's prop wash. Moran was spiraling toward the ground for a landing with his engine off when the accident happened. (Above, Edward Peck; below, Rick Tabb.)

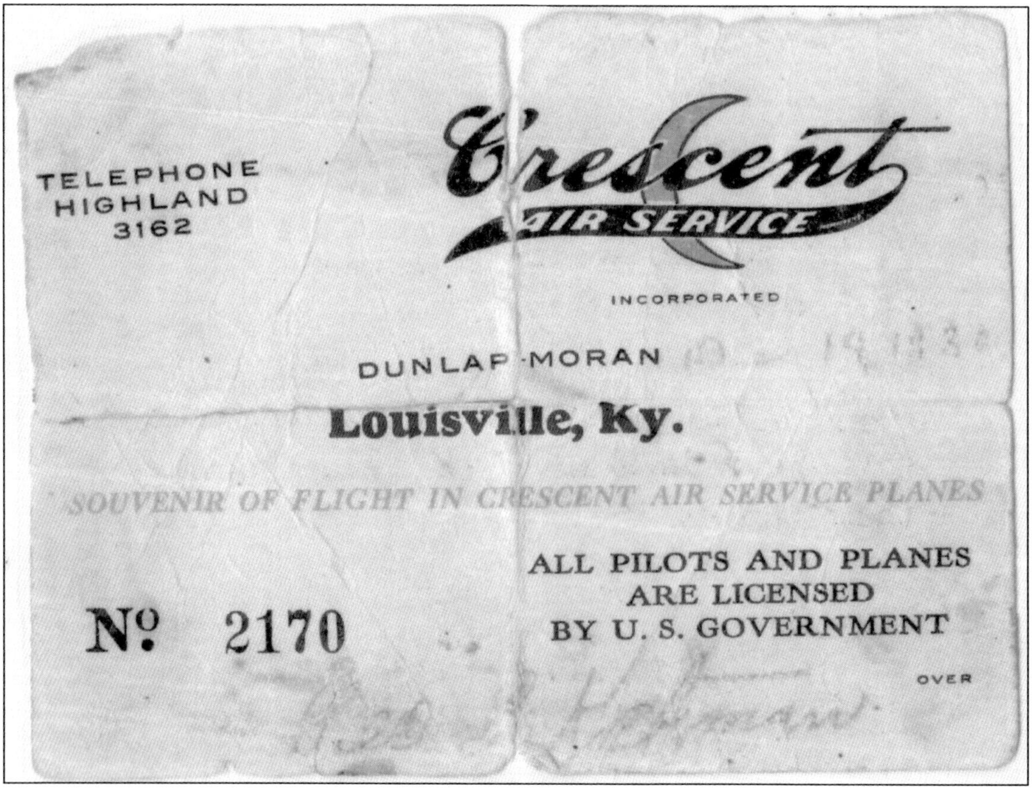

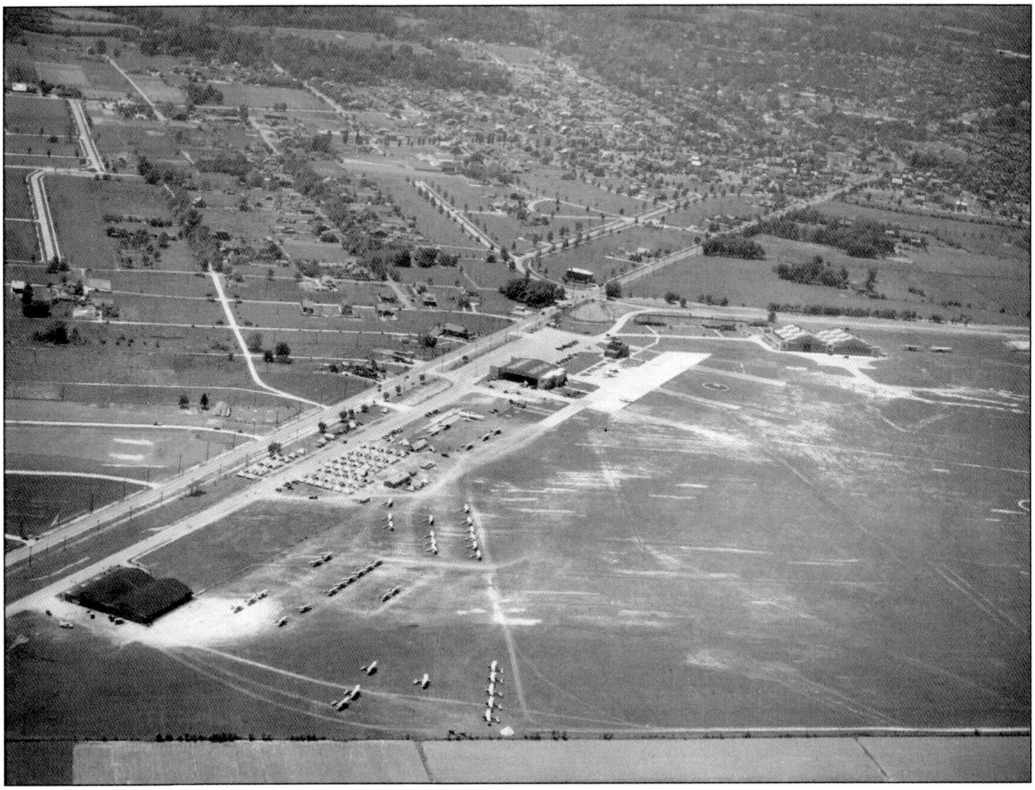

Bowman Field is clearly seen in this 1933 aerial view. The tent city between the hangars was erected by the 325th Observation Squadron Reserve to house visiting squadrons for mutual training exercises. They were referred to as "contact camps." This particular contact camp was for a war game known as the Joint Anti-Aircraft Exercise. At this time, Bowman Field supplied fighter aircraft to protect Fort Knox to the south from attacking bombers coming from the north. (Author's collection.)

The collapsed tent in this photograph is actually an officer's latrine setup for the 1933 Anti-Aircraft—Air Corps Joint Exercise. During this May exercise, four A-8 aircraft were attempting to land at Bowman Field during a rainstorm, with three of them going astray, plowing into tents in the bivouac area and the latrine tent. A portion of the airplane can be seen to the left of the tent. (Author's collection.)

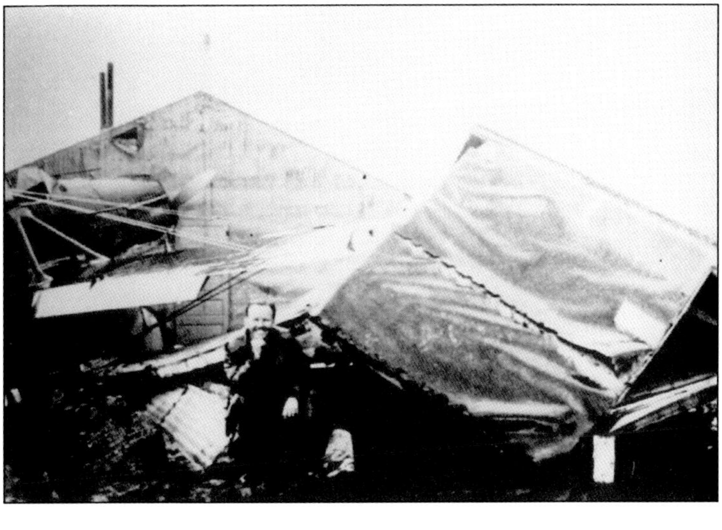

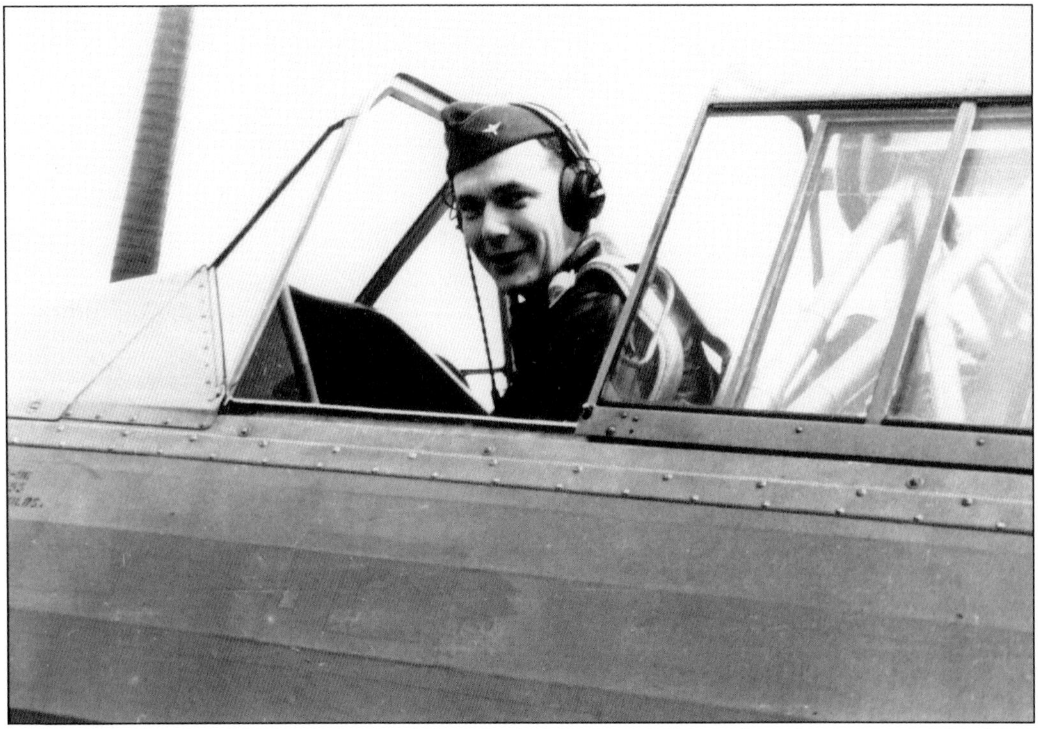

Pilot Ledcreich Stuart "L.S." Vance of the 325th Observation Squadron Reserve is in the front seat of a North American BT-9 trainer. Considering that reserve service amounted to a part-time job, Vance was also a member of the Louisville and Jefferson County Air Board, which was also part time. His regular job was an engineer with the Louisville Water Company. Vance was one of many individuals in Louisville dedicated to making Bowman Field a success. (Edward Peck.)

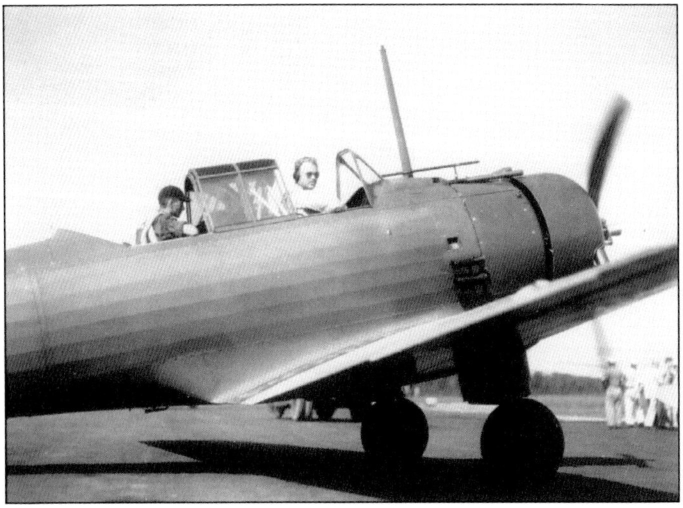

Norman Lewellyn commands a 325th Observation Squadron Reserve BT-9 with an observer in the back seat. Lewellyn was a Louisville businessman who went on active duty with the Army Air Forces at the beginning of World War II. He was killed in a B-17 bomber takeoff accident in India. The 325th was dissolved before the war, although many of its personnel transferred to active duty assignments during the war. (Robert Lewellyn.)

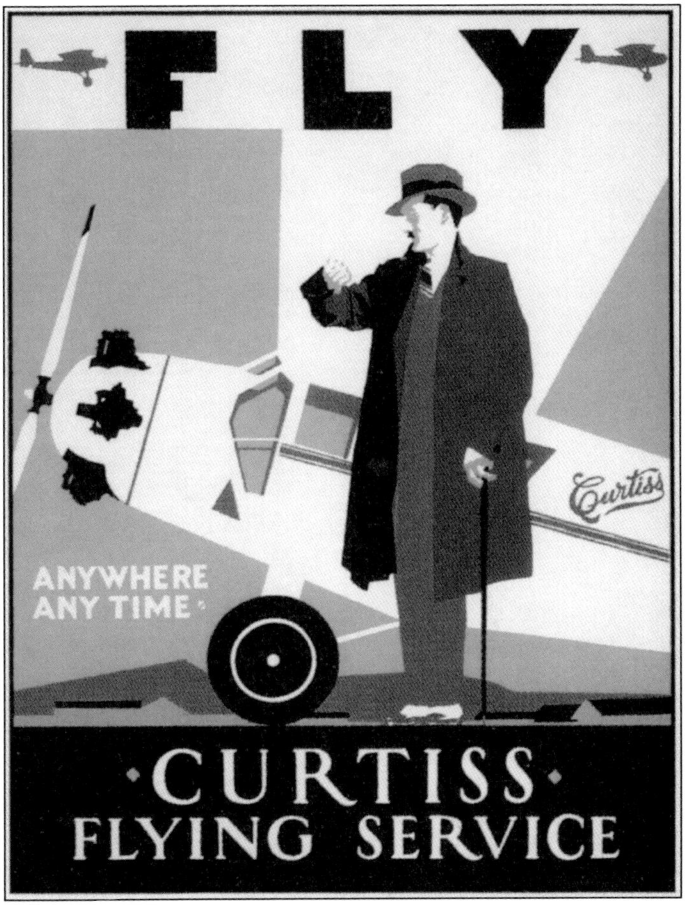

The Curtiss-Wright Aircraft Company was a diversified airplane manufacturer with factories across the country. But it was this St. Louis–produced Curtiss Robin that symbolized the Curtiss Flying Service division. It was used in the flying service's logo and advertising. Product recognition and identity were as important in the 1930s as today. For a pilot who wanted the best and to be seen in the best, it was the Robin. This was also true at Louisville's Curtiss Flying Service. (Right, author's collection; below, William Russell Beeler.)

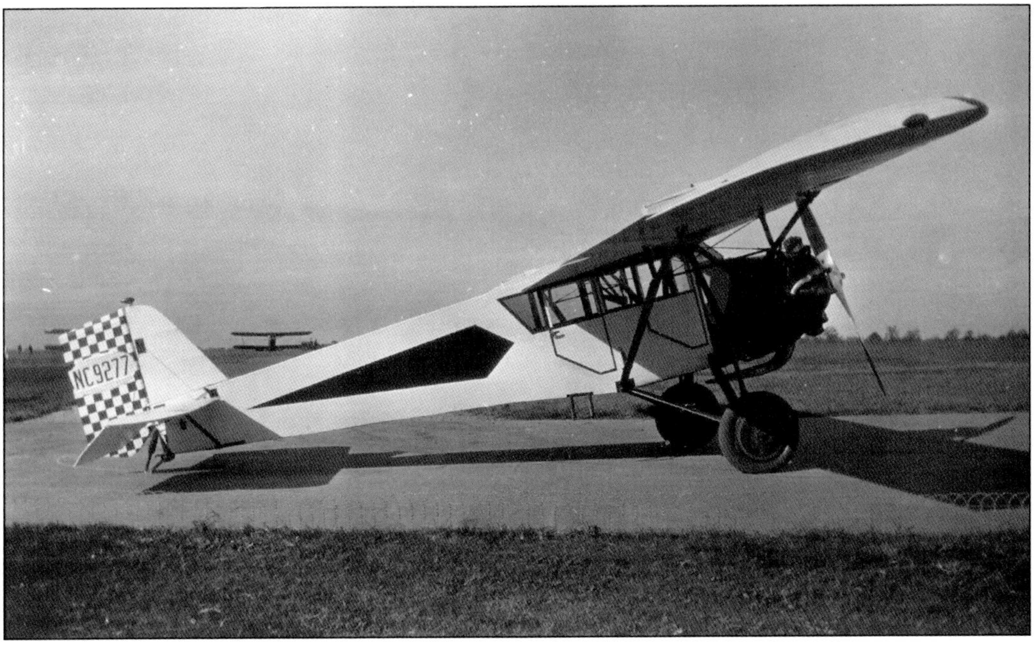

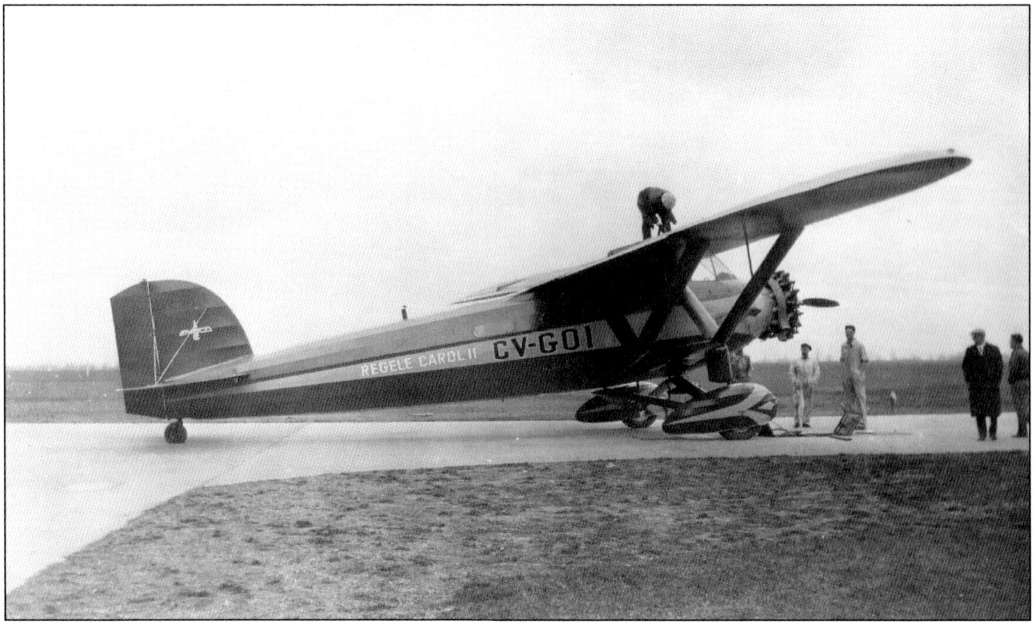

An Emsco B-3-A airplane said to be owned by a Romanian prince was at Bowman Field during February 1931. Curtiss-Wright Flying Service mechanics Frank Hand, Jimmy Wilson, and Frank Holzaphel tend to the airplane while the flight crew observe. It is not known if the prince accompanied the Emsco. (William Russell Beeler.)

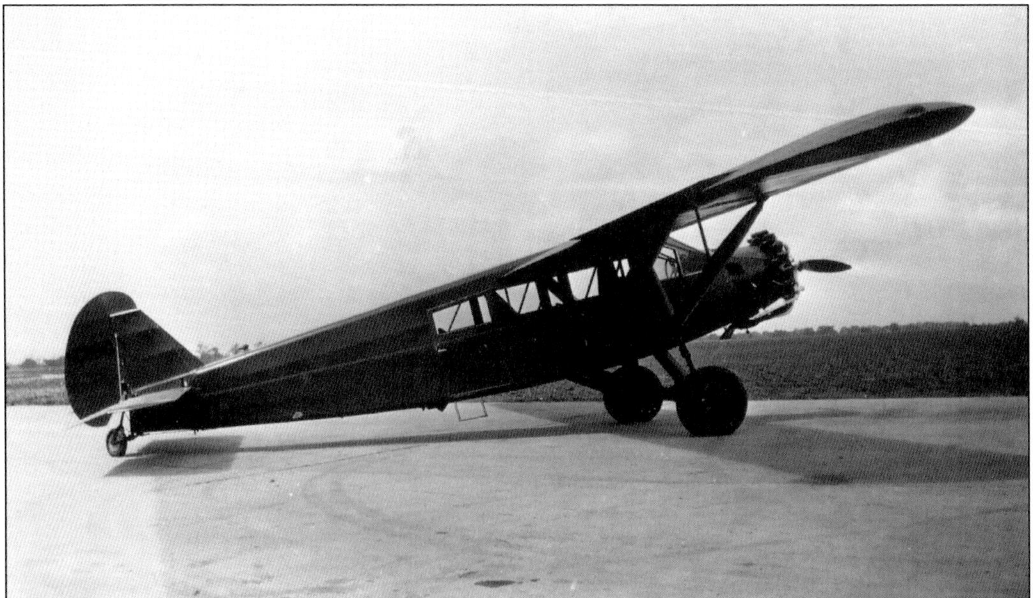
One of the aircraft owned by Curtiss Flying Service at Bowman Field was this TravelAir 6000. It arrived on April 16, 1930, and was used for passenger charter flights and sightseeing. The 1929 merger of the Curtiss and Wright aircraft companies saw this new entity purchase several other firms that made airplanes, including TravelAir. (William Russell Beeler.)

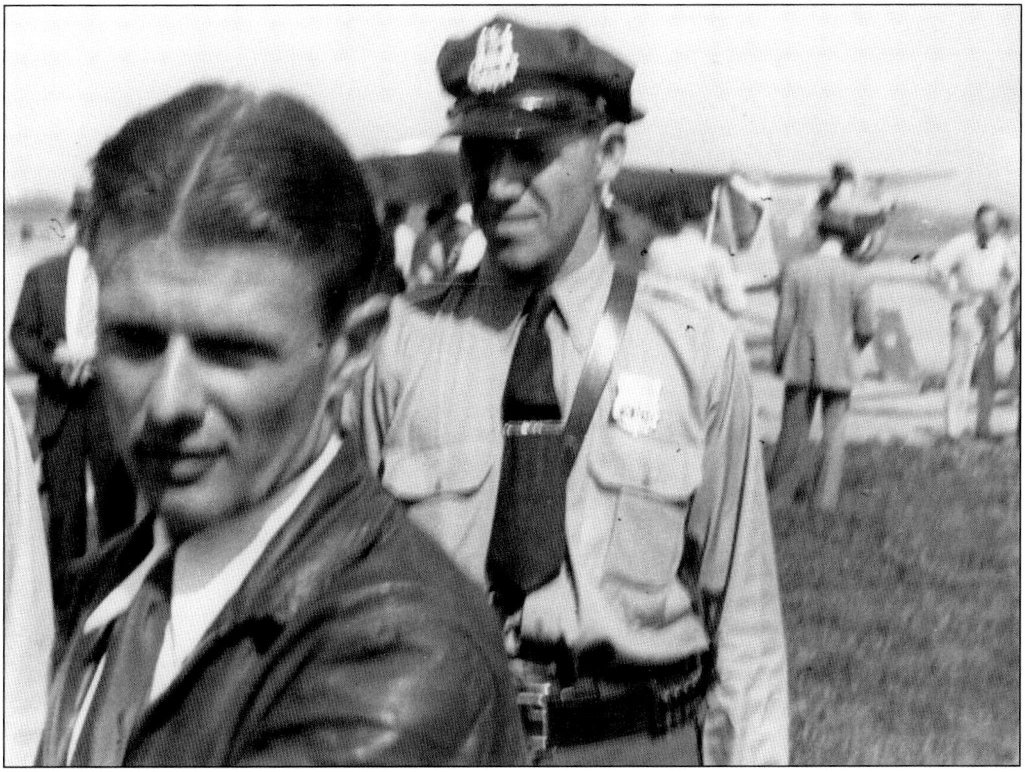

One of the most fascinating aviators of the 1930s was Douglas "Wrong Way" Corrigan, seen here at Bowman Field during his visit to Louisville on September 4, 1938. Corrigan got the name earlier in July when he took off from Long Island, New York, for a planned trip to California but instead flew to Dublin, Ireland. His explanation was that he took off in fog and lost his way. Corrigan said that he came to Louisville to enjoy the parties. (Edward Peck.)

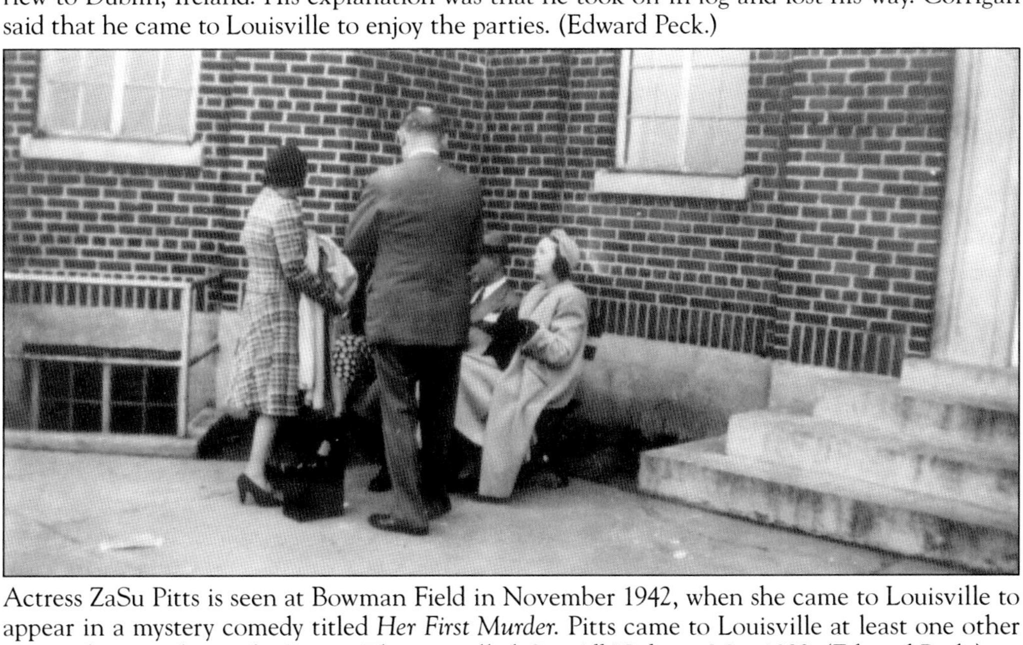

Actress ZaSu Pitts is seen at Bowman Field in November 1942, when she came to Louisville to appear in a mystery comedy titled *Her First Murder*. Pitts came to Louisville at least one other time to be in a play at the Brown Theater called *Out All Night*, in May 1933. (Edward Peck.)

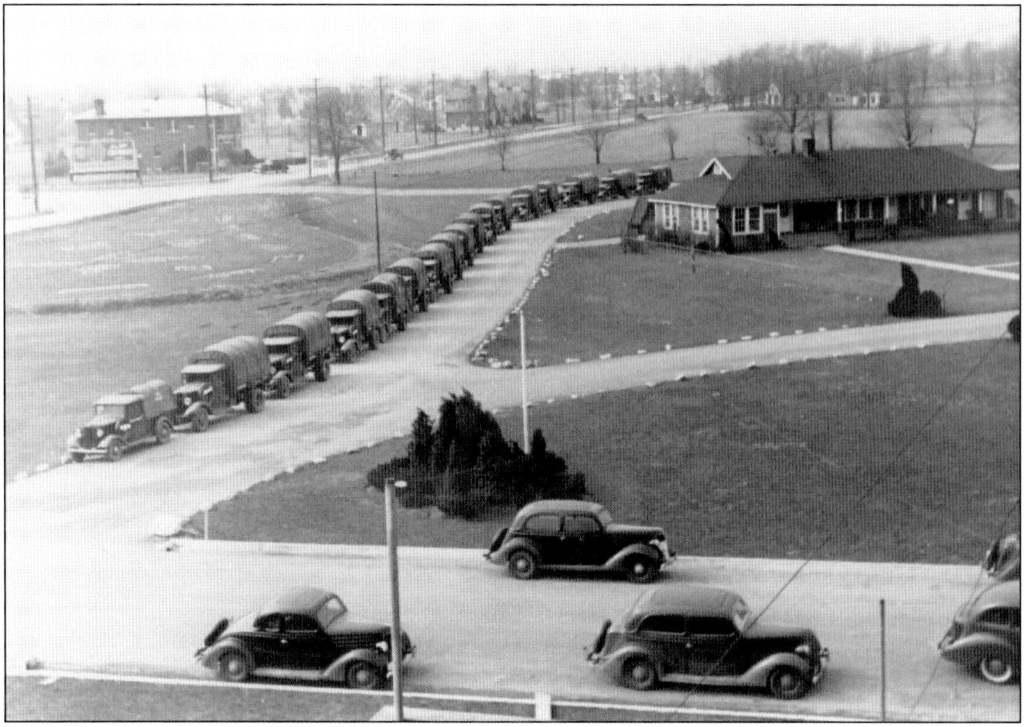

In January 1937, Louisville recorded nearly 20 inches of rain for the month resulting in the worst flood in the city's history. Flooding put 60 percent of Louisville under water and shut down the city's road and railroad system to the outside. Fortunately, Bowman Field continued to operate through the crisis. Although not having paved runways, Bowman Field's excellent drainage kept the airport open. When the ordeal of the flood was over, airlines were credited with bringing in 185,000 pounds of emergency supplies. They also brought doctors and nurses along with vital medications. Shown here is a busy Bowman Field during the flood. Military trucks await supplies airlifted in and extra personnel are on duty to handle it. (Both, Edward Peck.)

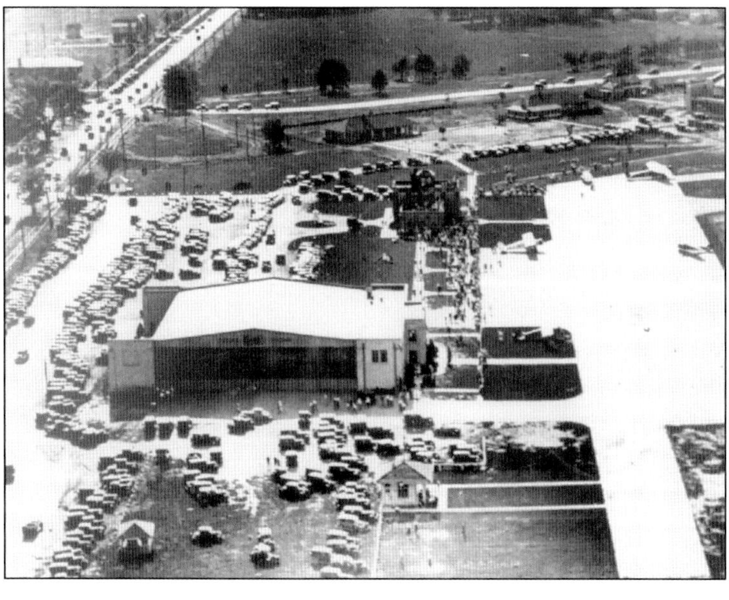

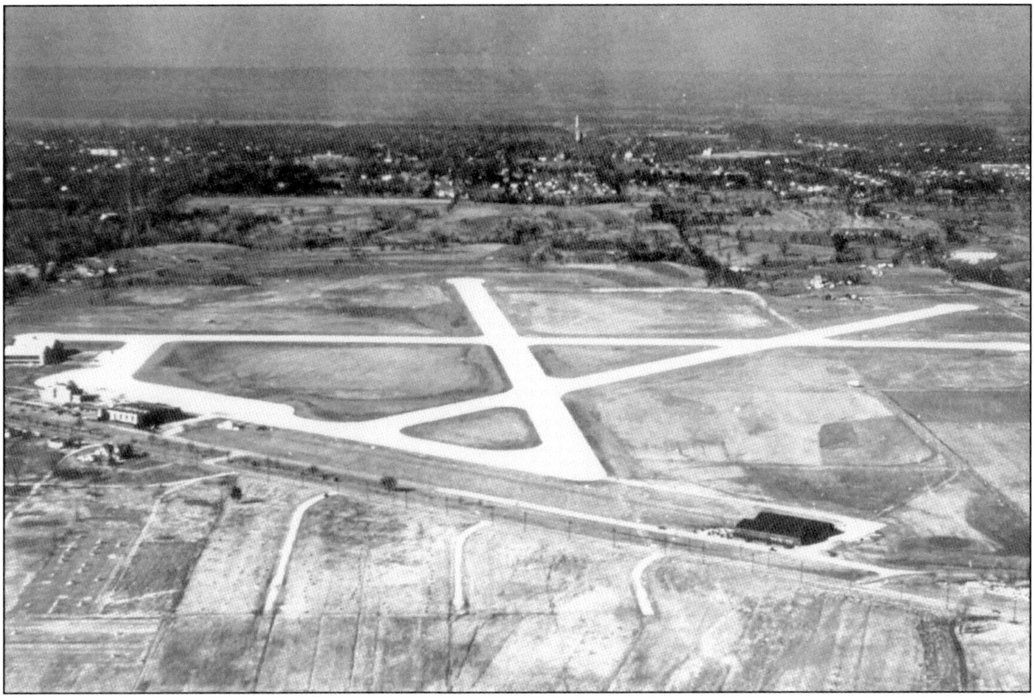

As a result of the 1937 flood, it was determined that Bowman Field needed paved runways. Excavation and preparation for three new paved runways started early in 1938 and continued for much of the year to make Bowman Field truly an all-weather airport. When the first concrete was poured in April, Abram Bowman (center right below) gathered with other dignitaries for the ceremonial release. These runways would serve well into the future, although one would be removed during the 1990s as a cost saving measure. (Above, RAA; below, Ed Taylor.)

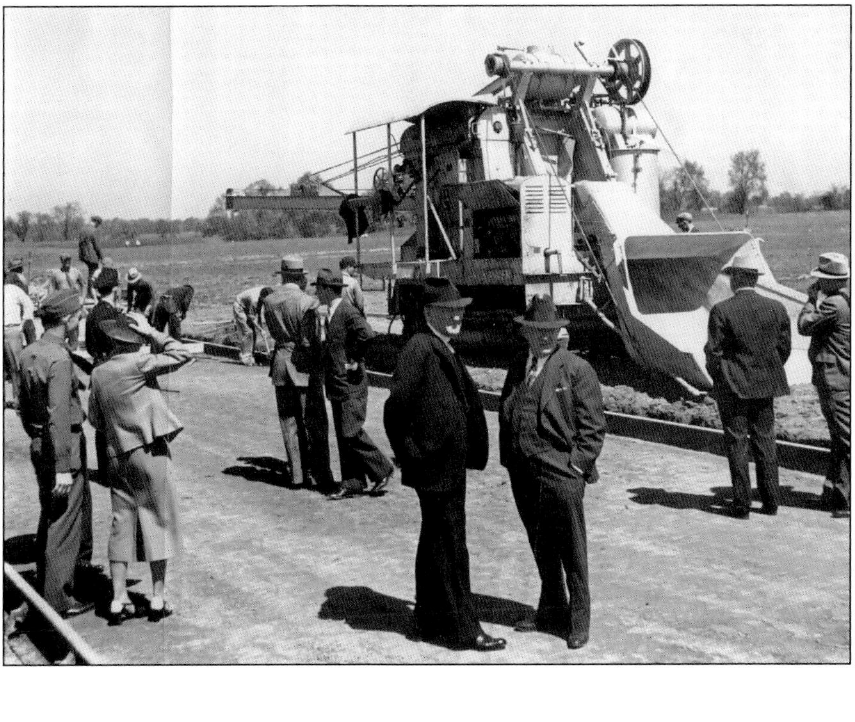

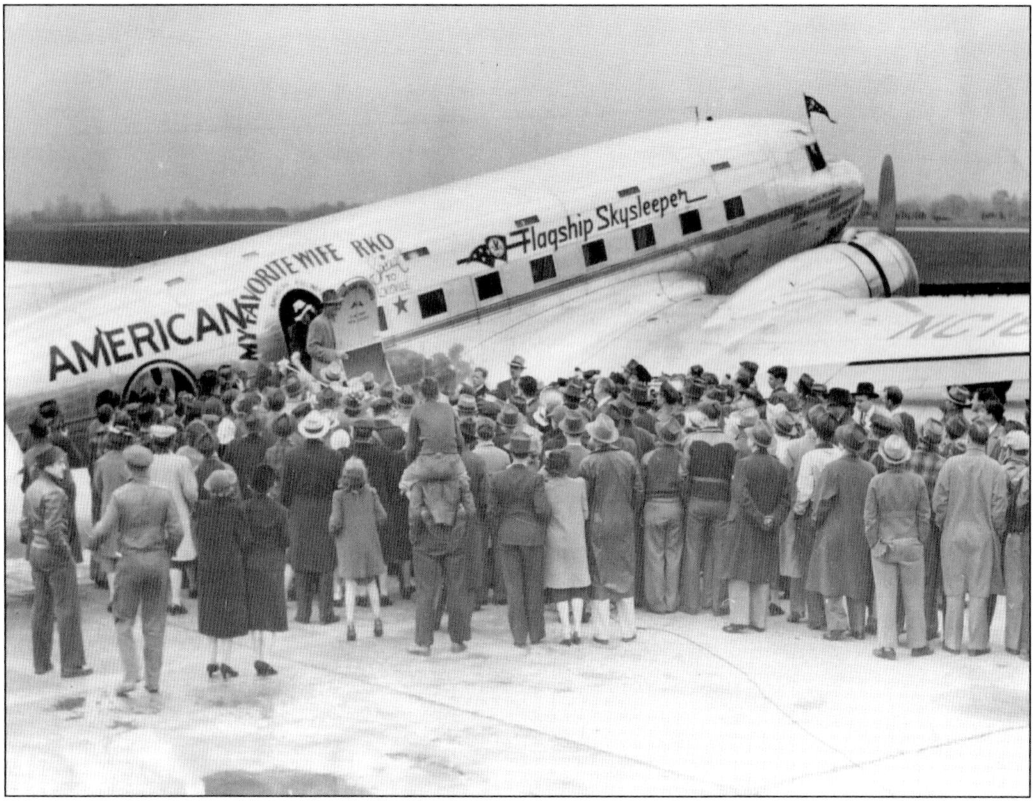

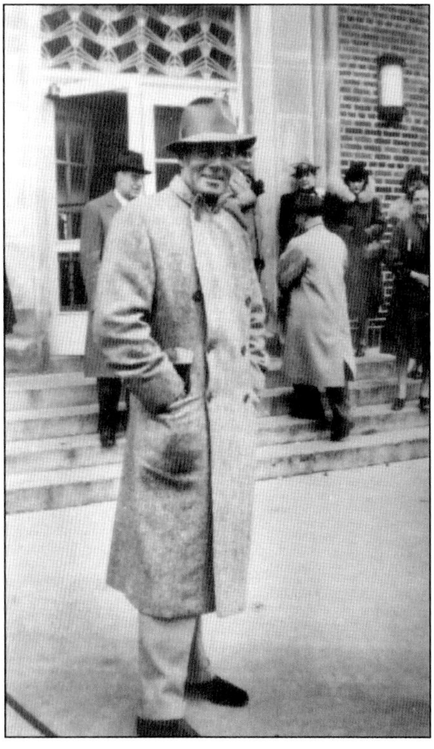

Movie star and Louisville native Irene Dunn is shown arriving at Bowman Field aboard an American Airlines DC-3 on May 8, 1940. She was in town promoting her new movie, *My Favorite Wife*, with costar Randolph Scott. Both were invited to Louisville by Kentucky governor Keen Johnson for the 66th Kentucky Derby. Scott was later photographed by the Bowman Field administration building. (Above, author's collection; left, Edward Peck.)

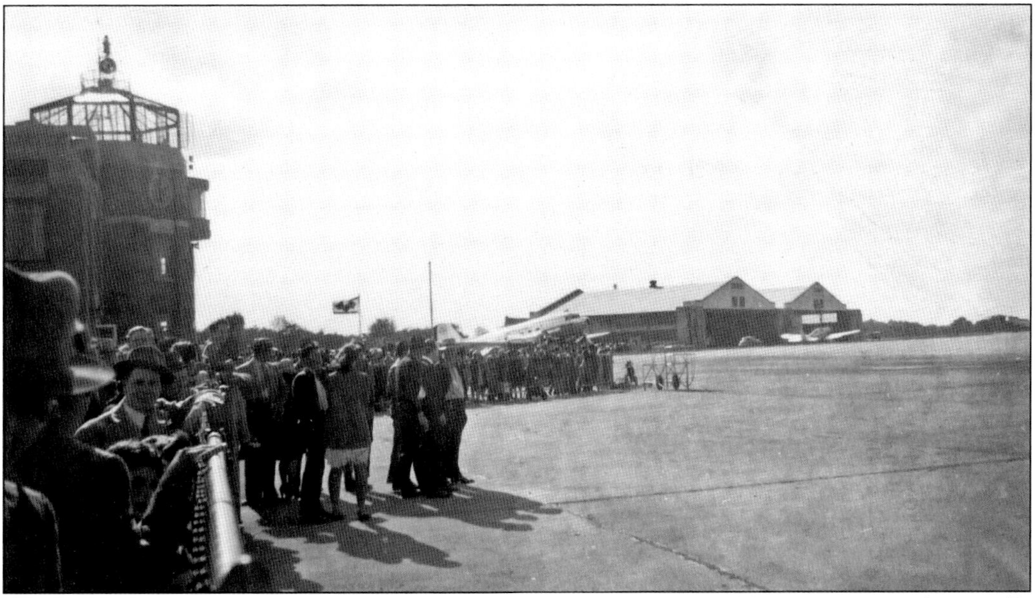

The effects of the Great Depression during the 1930s brought many hardships to the city of Louisville and the nation as well. With many out of work, money was in short supply and sacrifices had to be made. One way to solve this was to look for free Depression-era entertainment. This brought flocks of people to Bowman Field, especially on the weekends, to watch airplanes. Checking airline schedules in the newspaper, one could be at the field at the appropriate time to catch the most flights. And there was always the chance to see a celebrity, famous person, or sports star board or depart an airplane. (Above, Edward Peck; below, RAA.)

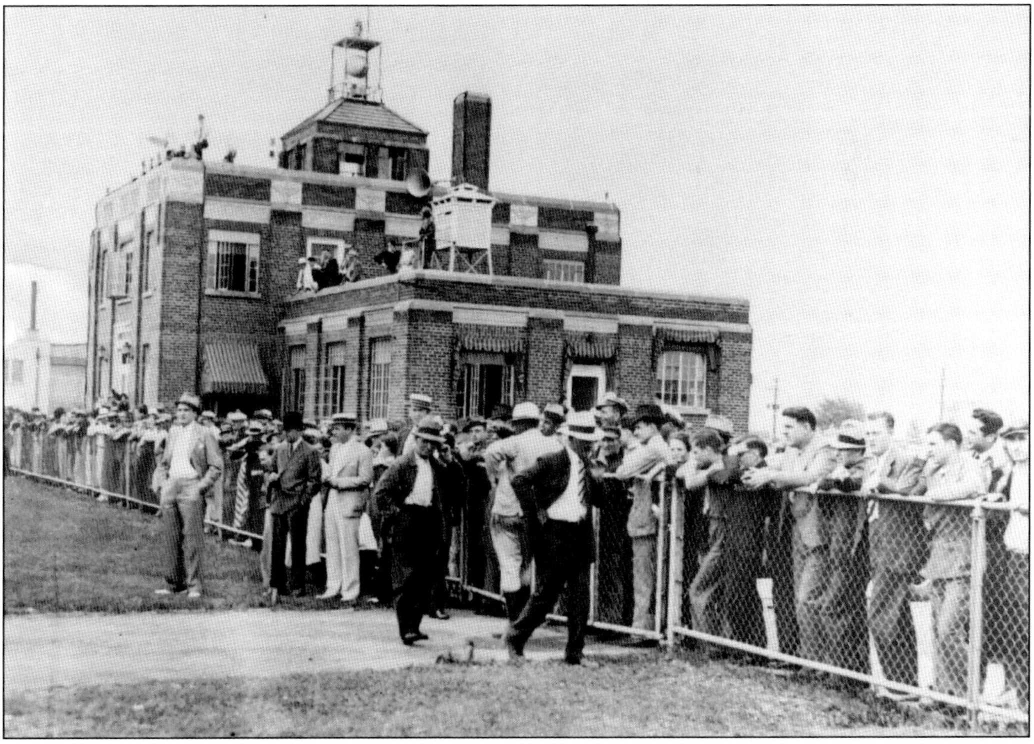

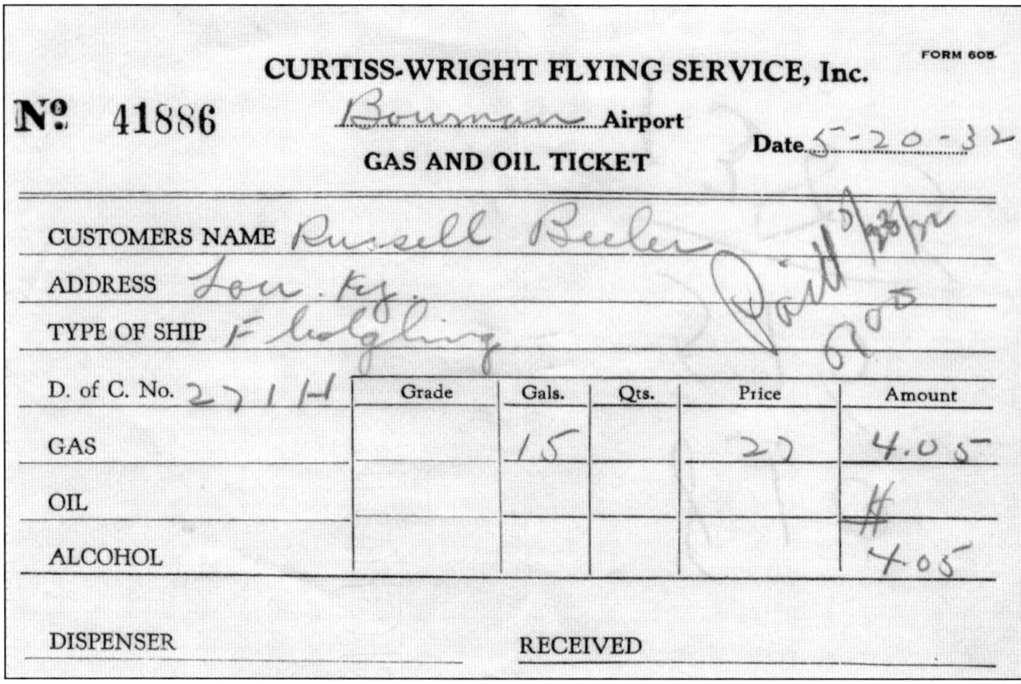

CURTISS-WRIGHT FLYING SERVICE, Inc. FORM 605

№ 41886 Bowman Airport

GAS AND OIL TICKET

Date 5-20-32

CUSTOMERS NAME Russell Beeler

ADDRESS Lou. Ky.

TYPE OF SHIP Fledgling

D. of C. No. 271H	Grade	Gals.	Qts.	Price	Amount
GAS		15		27	4.05
OIL					#
ALCOHOL					4.05

DISPENSER RECEIVED

During the 1930s, Curtiss Flying Service, which changed its name to Curtiss-Wright Flying Service and then to Louisville Flying Service, was the primary place at Bowman Field for service and gasoline. Here are two gasoline tickets from 1932 and 1933 that show that the price dropped 2¢ during the latter year to 25¢ per gallon. The purchaser was Russell Beeler from New Albany, who owned the Jeffersonville Airport, which opened in 1930. (Both, William Russell Beeler.)

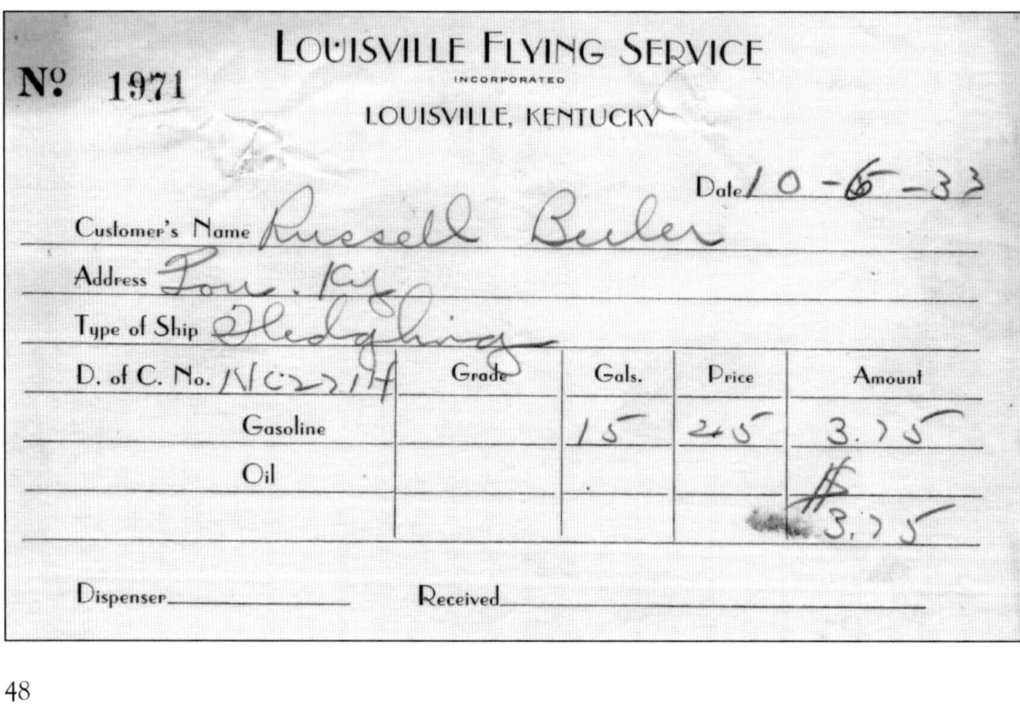

LOUISVILLE FLYING SERVICE

INCORPORATED

LOUISVILLE, KENTUCKY

№ 1971

Date 10-6-33

Customer's Name Russell Beeler

Address Lou. Ky.

Type of Ship Fledgling

D. of C. No. NC271H	Grade	Gals.	Price	Amount
Gasoline		15	25	3.75
Oil				#
				3.75

Dispenser Received

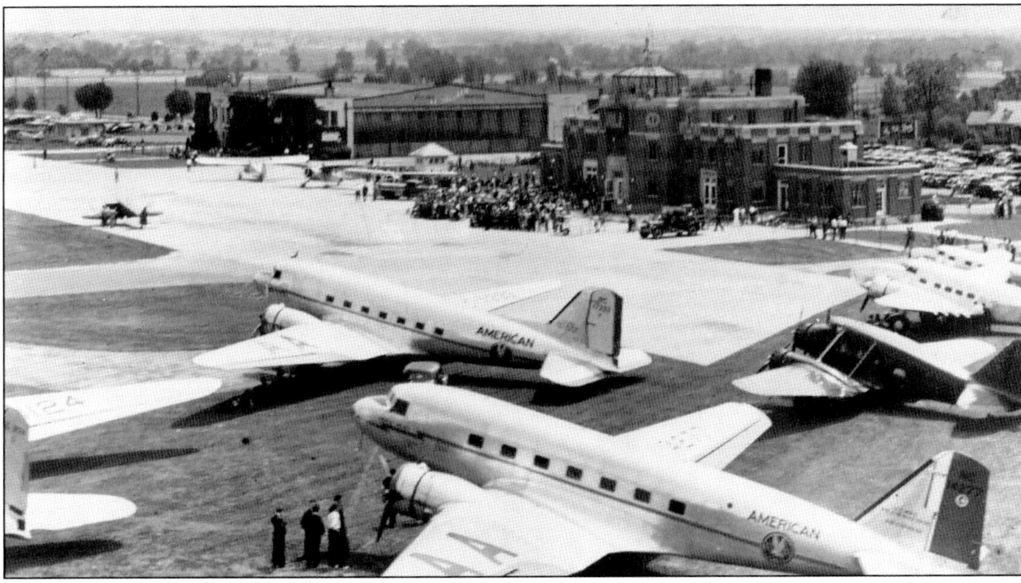

The first Saturday in May has always been a special day in Louisville. The Kentucky Derby draws numerous visitors to the city each year. When the airlines served Bowman Field, it became a natural spot to gather with hopes of seeing a celebrity. To accommodate derby visitors, both American and Eastern Air Lines would add extra flights, which resulted in more airplanes at Bowman Field. These two images show just that—plenty of people and airplanes at derby time. (Both, RAA.)

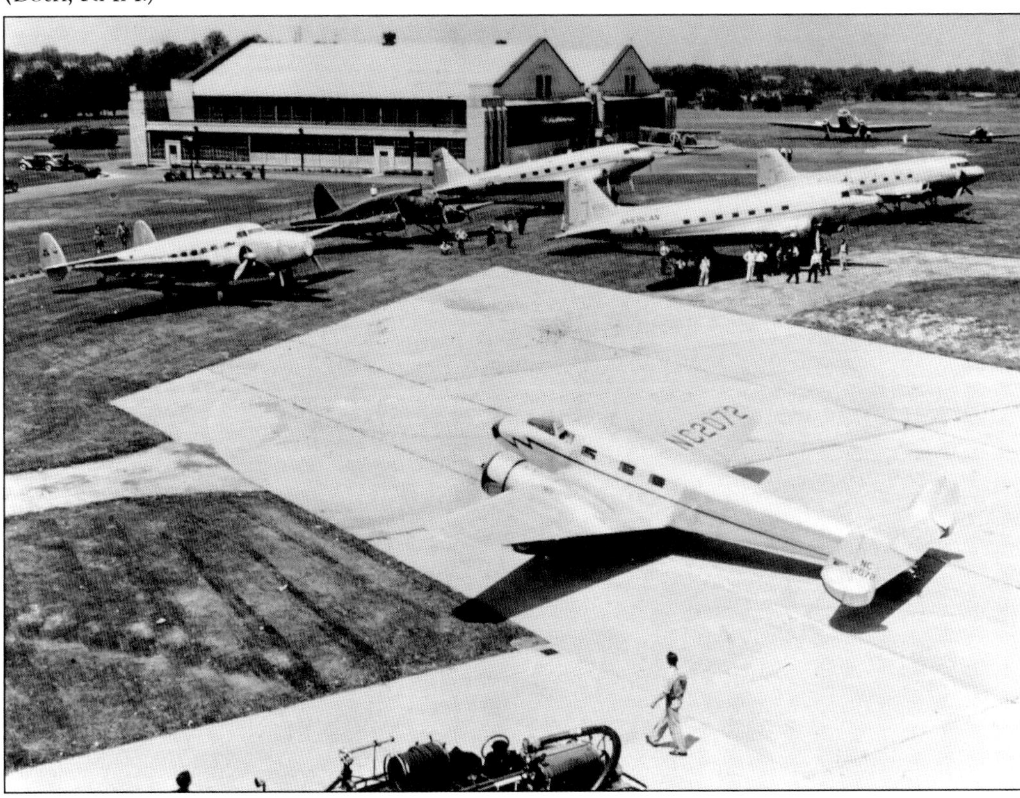

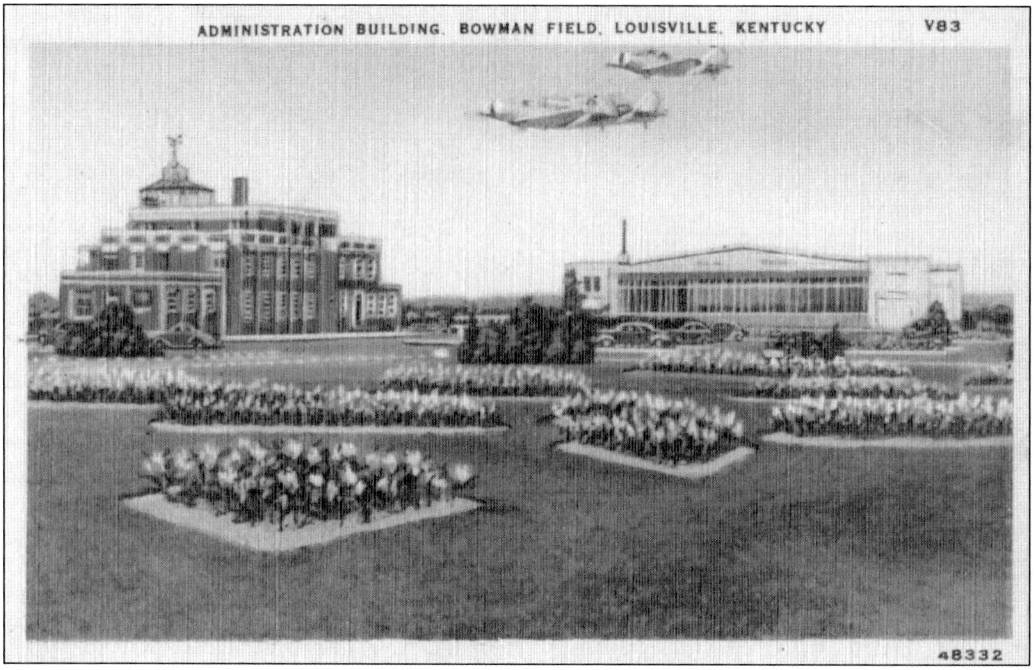

Postcards have long been a popular way to communicate and send greetings, and the popularity of a location could be determined by how many different postcards were available. Bowman Field had numerous postcards, mostly during the 1930s. Bowman Field postcards were usually colorized black-and-white photographs that were sometimes enhanced with artist-added features such as airplanes in flight. Bowman Field postcards were a combination of aerial and ground views of the airport. (Both, author's collection.)

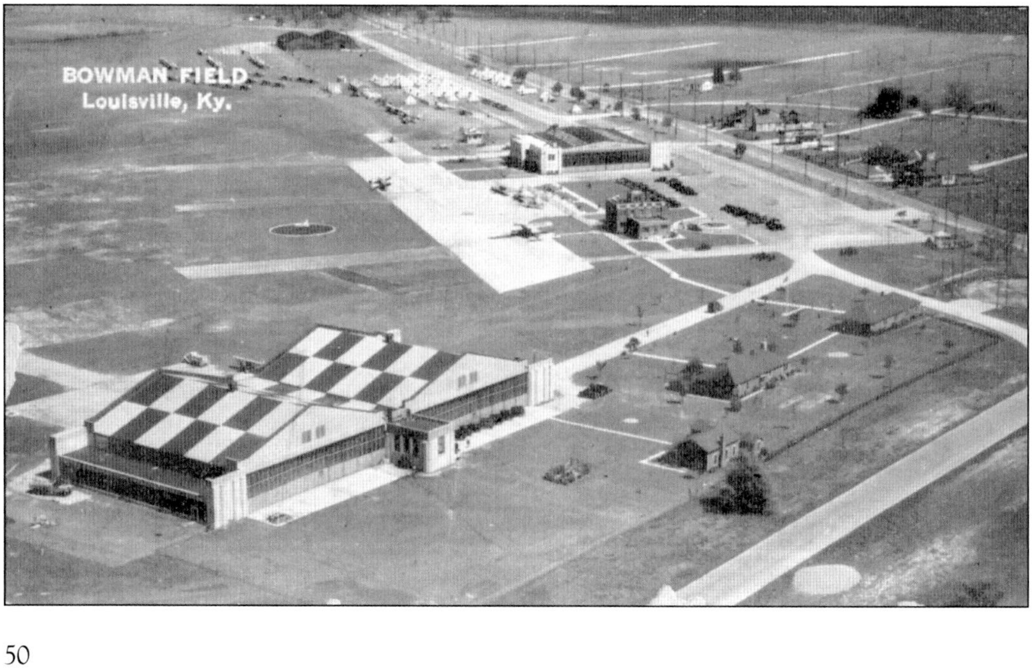

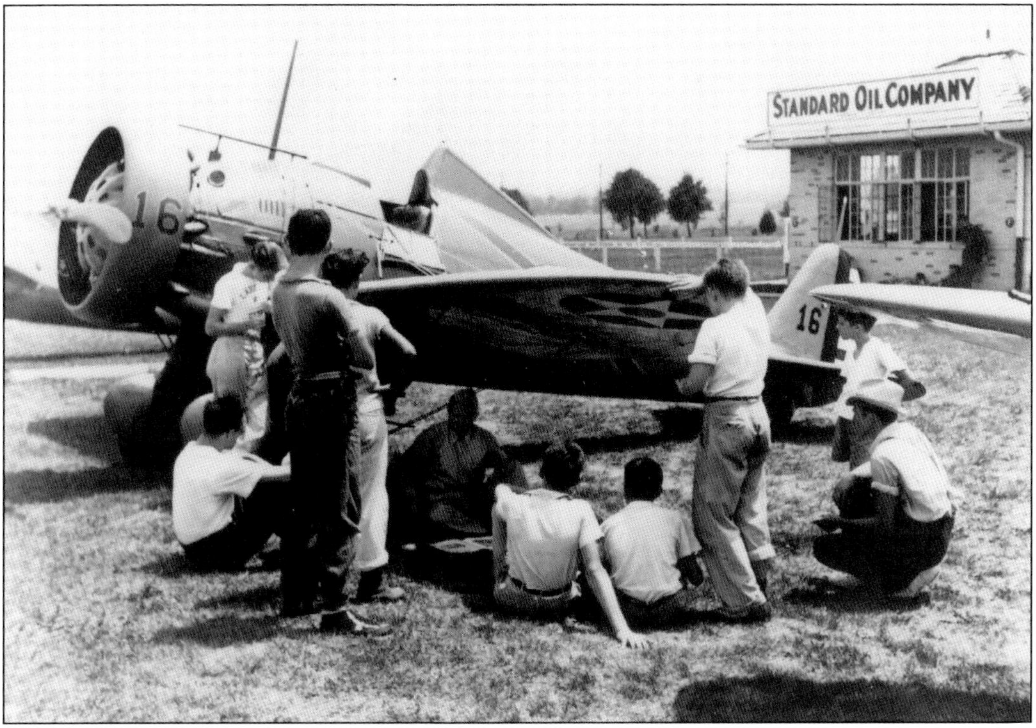

Nothing exemplifies the saying "boys will be boys" more than these two images of a P-26 fighter at Bowman Field. Young boys are charmed by the pilot while they get a close look at the dream plane of the day. The P-26, nicknamed "Peashooter," was the neatest airplane ever in the mind of a future aviator. It was at Bowman Field during the 1930s for an air show or display. The Standard Oil building in the background was located behind the Louisville Flying Service hangar and provided gasoline pumps for aircraft. (Both, Edward Peck.)

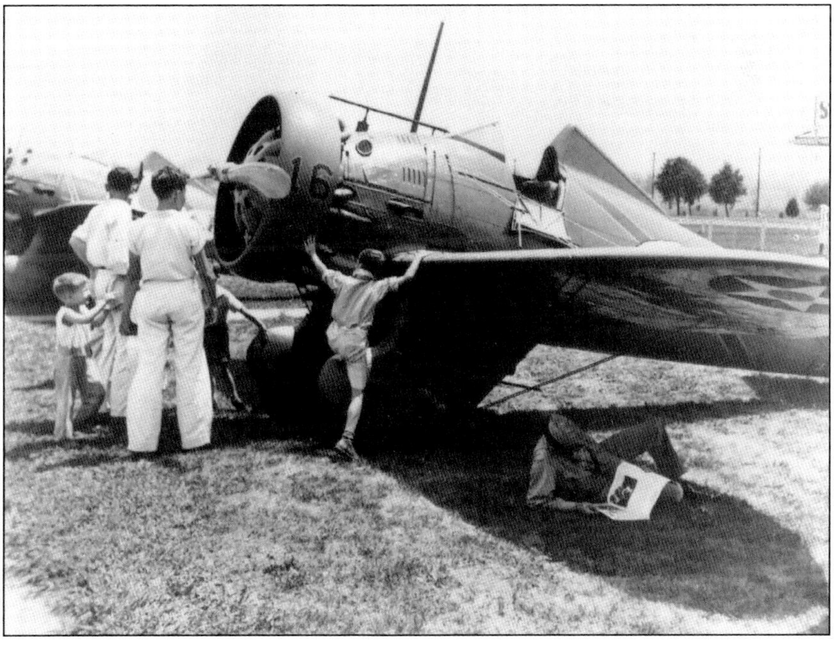

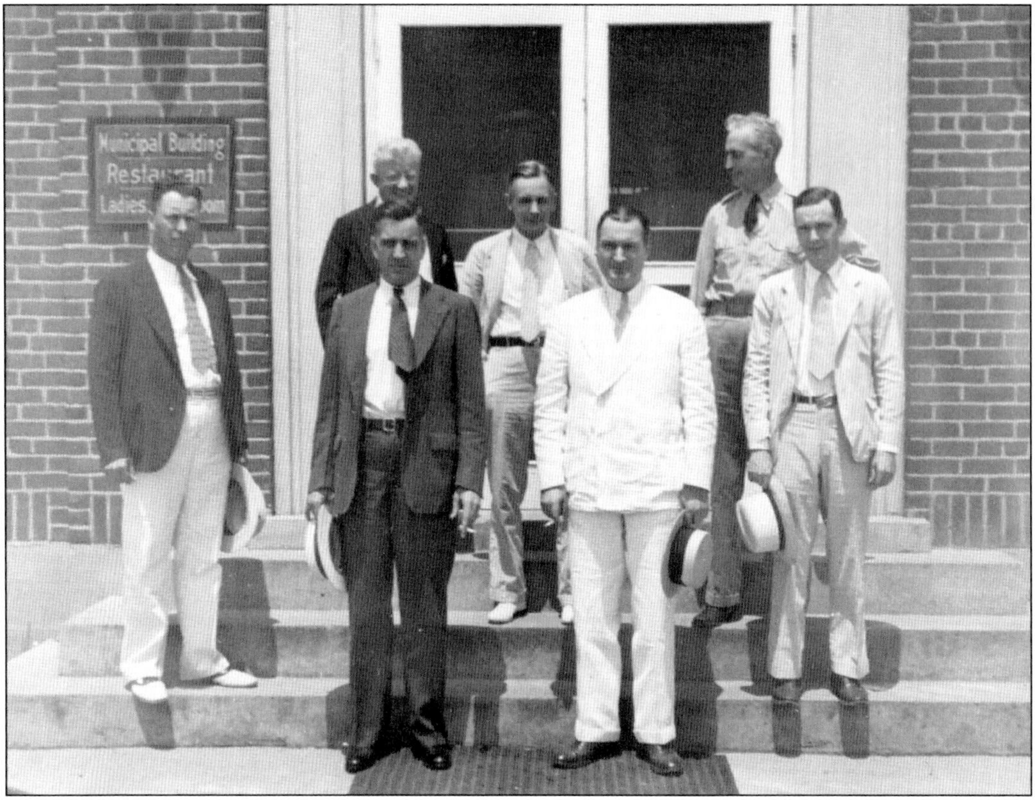

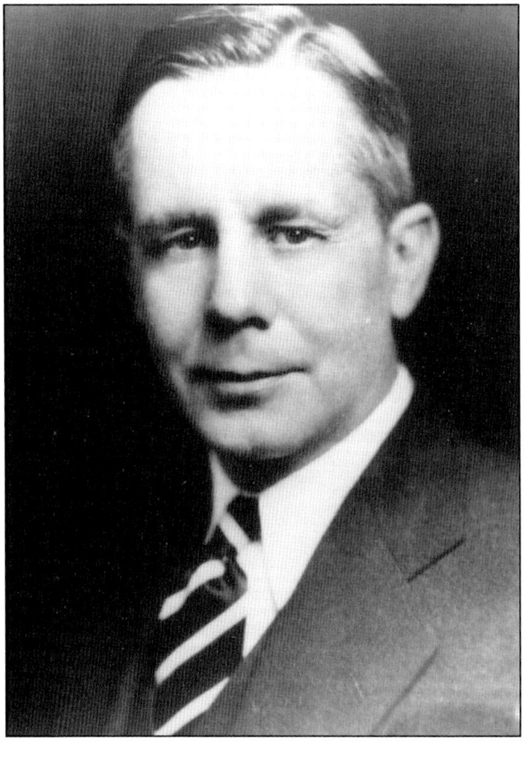

Members of the first Louisville and Jefferson County Air Board pose for a photograph on the steps of the administration building. From left to right are (first row) Al Near (under sign), Al Woody, David Farleigh, and L.S. Vance; (second row) Roger Schupp (secretary), Addison Lee (chairman), and unidentified. (Edward Peck.)

It takes great people to make a great airport. Addison W. Lee Jr. was the first chairman of the Louisville and Jefferson County Air Board, holding that position from 1928 until his death in 1949. He was an early advocate for a municipal airport in Louisville and worked to bring airlines and improved air service to Bowman Field. Trained as an electrical engineer, he was vice president of Louisville Gas and Electric at the time of his passing. He was also Abram Bowman's brother-in-law. (Edward Peck.)

Three

World War II

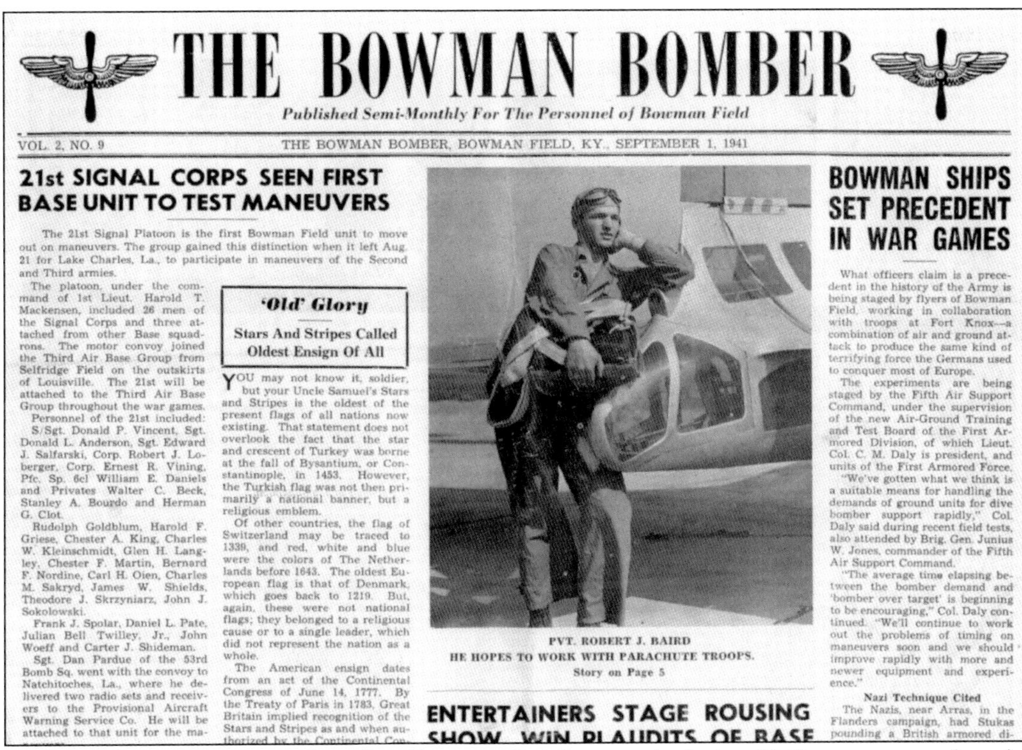

THE BOWMAN BOMBER

Published Semi-Monthly For The Personnel of Bowman Field

VOL. 2, NO. 9 — THE BOWMAN BOMBER, BOWMAN FIELD, KY., SEPTEMBER 1, 1941

21st SIGNAL CORPS SEEN FIRST BASE UNIT TO TEST MANEUVERS

The 21st Signal Platoon is the first Bowman Field unit to move out on maneuvers. The group gained this distinction when it left Aug. 21 for Lake Charles, La., to participate in maneuvers of the Second and Third armies.

The platoon, under the command of 1st Lieut. Harold T. Mackensen, included 26 men of the Signal Corps and three attached from other Base squadrons. The motor convoy joined the Third Air Base Group from Selfridge Field on the outskirts of Louisville. The 21st will be attached to the Third Air Base Group throughout the war games.

Personnel of the 21st included:

S/Sgt. Donald P. Vincent, Sgt. Donald L. Anderson, Sgt. Edward J. Salfarski, Corp. Robert J. Loberger, Corp. Ernest R. Vining, Pfc. Sp. 6cl William E. Daniels and Privates Walter C. Beck, Stanley A. Bourdo and Herman G. Clot.

Rudolph Goldblum, Harold F. Griese, Chester A. King, Charles W. Kleinschmidt, Glen H. Langley, Chester F. Martin, Bernard F. Nordine, Carl H. Oien, Charles M. Sakryd, James W. Shields, Theodore J. Skrzyniarz, John J. Sokolowski.

Frank J. Spolar, Daniel L. Pate, Julian Bell Twilley, Jr., John Woeff and Carter J. Shideman.

Sgt. Dan Pardue of the 53rd Bomb Sq. went with the convoy to Natchitoches, La., where he delivered two radio sets and receivers to the Provisional Aircraft Warning Service Co. He will be attached to that unit for the ma-

'Old' Glory

Stars And Stripes Called Oldest Ensign Of All

YOU may not know it, soldier, but your Uncle Samuel's Stars and Stripes is the oldest of the present flags of all nations now existing. That statement does not overlook the fact that the star and crescent of Turkey was borne at the fall of Bysantium, or Constantinople, in 1453. However, the Turkish flag was not then primarily a national banner, but a religious emblem.

Of other countries, the flag of Switzerland may be traced to 1339, and red, white and blue were the colors of The Netherlands before 1643. The oldest European flag is that of Denmark, which goes back to 1219. But, again, these were not national flags; they belonged to a religious cause or to a single leader, which did not represent the nation as a whole.

The American ensign dates from an act of the Continental Congress of June 14, 1777. By the Treaty of Paris in 1783, Great Britain implied recognition of the Stars and Stripes as and when au-

PVT. ROBERT J. BAIRD
HE HOPES TO WORK WITH PARACHUTE TROOPS.
Story on Page 5

ENTERTAINERS STAGE ROUSING SHOW WIN PLAUDITS OF BASE

BOWMAN SHIPS SET PRECEDENT IN WAR GAMES

What officers claim is a precedent in the history of the Army is being staged by flyers of Bowman Field, working in collaboration with troops at Fort Knox—a combination of air and ground attack to produce the same kind of terrifying force the Germans used to conquer most of Europe.

The experiments are being staged by the Fifth Air Support Command, under the supervision of the new Air-Ground Training and Test Board of the First Armored Division, of which Lieut. Col. C. M. Daly is president, and units of the First Armored Force.

"We've gotten what we think is a suitable means for handling the demands of ground units for dive bomber support rapidly," Col. Daly said during recent field tests, also attended by Brig. Gen. Junius W. Jones, commander of the Fifth Air Support Command.

"The average time elapsing between the bomber demand and 'bomber over target' is beginning to be encouraging," Col. Daly continued. "We'll continue to work out the problems of timing on maneuvers soon and we should improve rapidly with more and newer equipment and experience."

Nazi Technique Cited

The Nazis, near Arras, in the Flanders campaign, had Stukas pounding a British armored di-

Additional acreage was purchased on Bowman Field's east side during 1940 to accommodate the expansion of the airport into the new Bowman Army Air Base. This added space was needed for the 46th Bomb Group and its four squadrons of Douglas A-20 light attack bombers. The military Bowman Field became an air base city with thousands of soldiers to keep the airplanes flying. Maintenance shops, mess halls, barracks, and a hospital were built over a period of several years. As with any city, a newspaper was important for the residents, and the *Bowman Bomber* filled that need. The *Bomber* was a professionally produced newspaper with sports, entertainment news, and advertising. (Author's collection.)

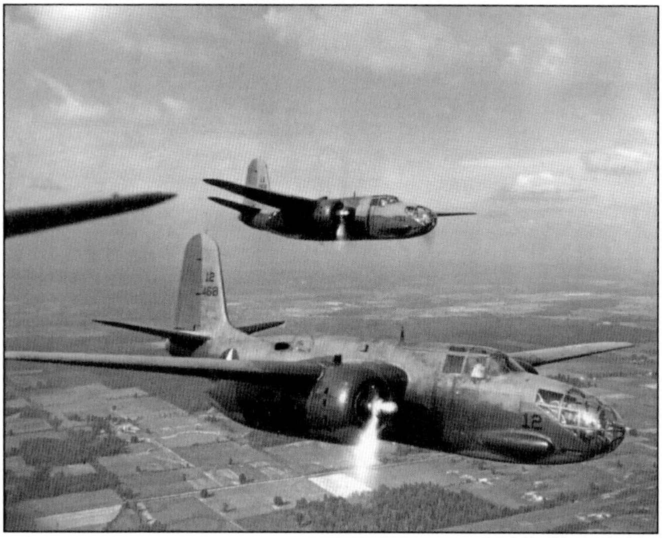

With "46B" indicating the 46th Bombardment Group painted on their tails, aircraft 11 and 12 fly over rural Kentucky for photographers. These are Douglas A-20 light bombers that served at the Bowman Army Air Base between 1941 and 1942. They were designed for low-level bombing missions. A target range was established at nearby Fort Knox for training. (Edward Peck.)

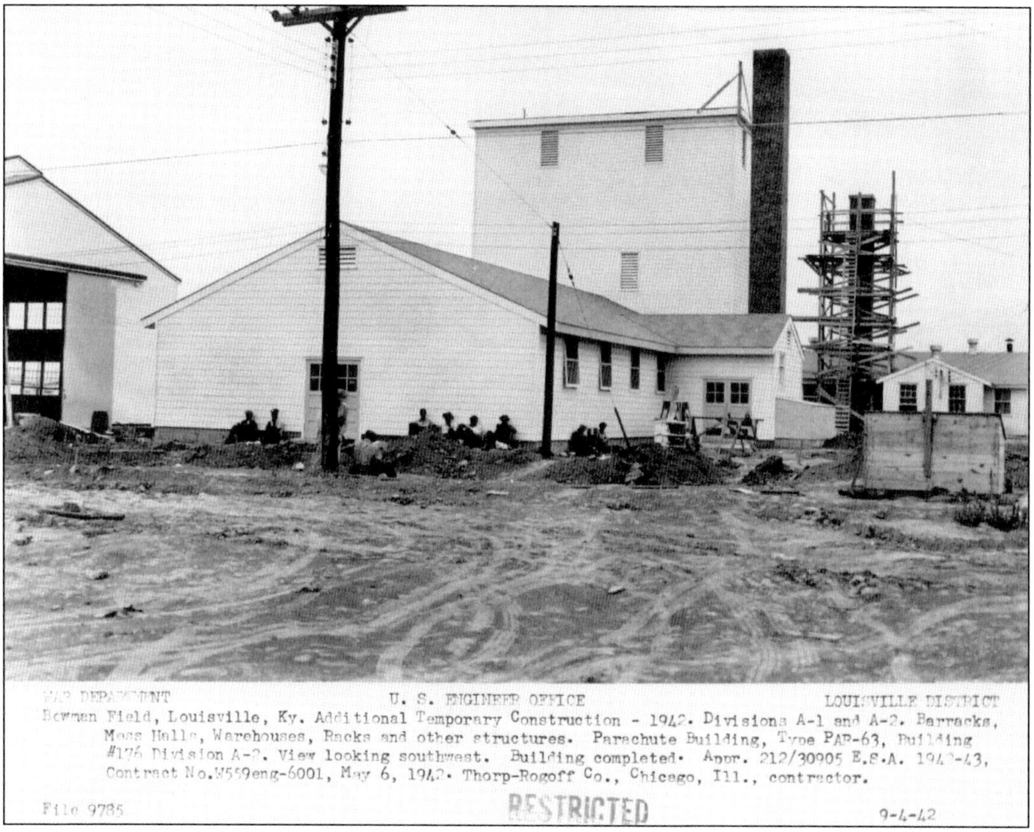

Construction of buildings for the new Bowman Army Air Base started in 1940. About 124 buildings of all sizes were completed, including everything from one large hangar to small guard shacks and everything in between. This photograph from September 4, 1942, shows the parachute building near completion. Since parachutes have to be inspected periodically, they must be unpacked and hung for a period of time before being repacked, which explains the large loft seen in the background. (COE.)

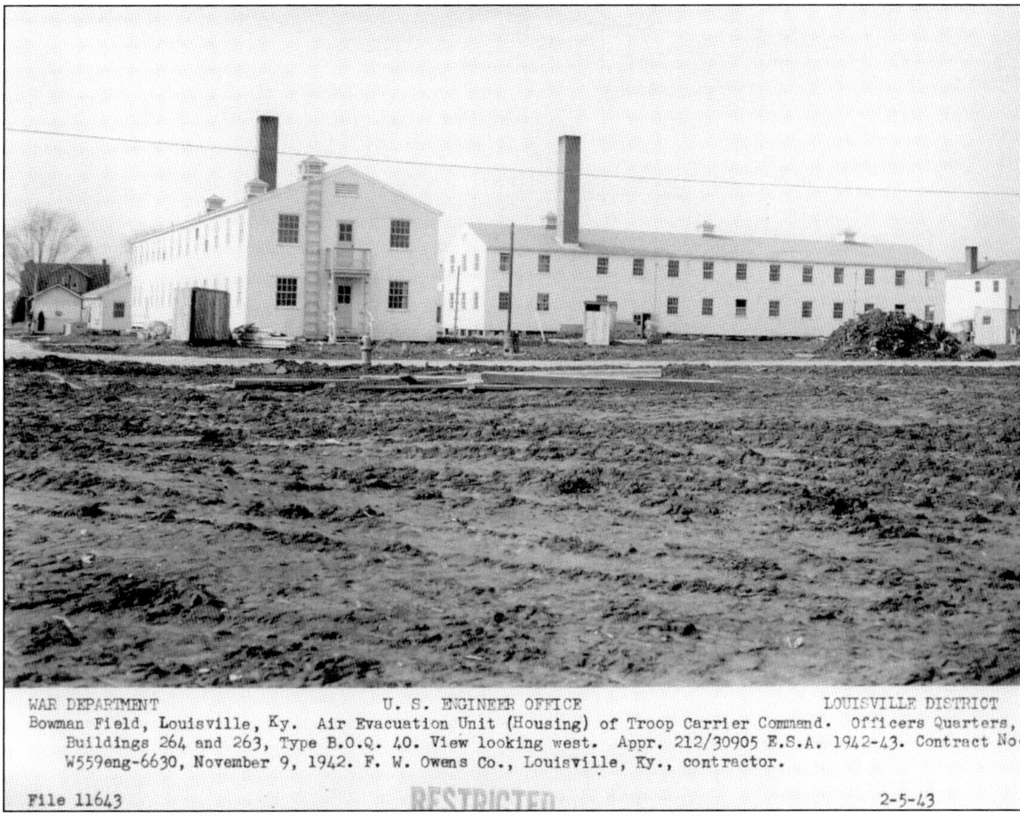

With the arrival of the School of Aerial Evacuation at the Bowman Army Air Base in 1942, quarters for officers had to be expanded. In this February 5, 1943, photograph, an area has been cleared for construction of additional Bachelor Officers Quarter, or BOQs. Since building construction at the Bowman Army Air Base fell under the classification of temporary construction, only for the duration of the war, most buildings were made with wood. They would be gradually demolished in the years after the war. The last one was the old clubhouse of the Aero Club. (COE.)

Even with all the airplanes at the Bowman Army Air Base during World War II, only one new hangar was built. This large hangar, when completed, had a black roof and was referred to as the "Black Hangar." Perhaps the black roof set the hangar apart from the many white buildings on base, giving incoming airplanes an idea where hangar space and aircraft service was available. This building is the only wartime constructed building that has survived on the field today. For years, it was used by Kentucky Flying Service and currently is used by Louisville Executive Aviation. (COE.)

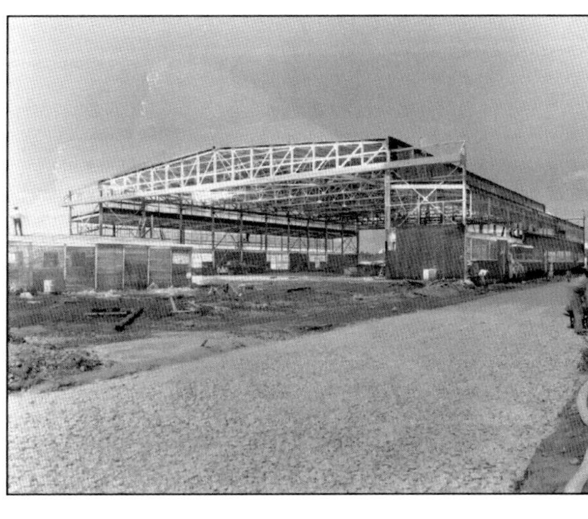

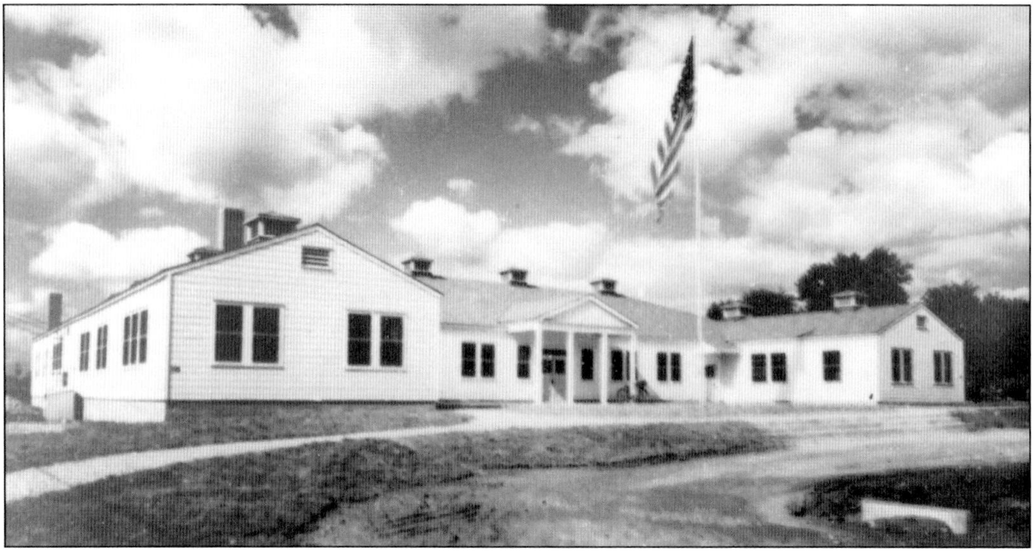

Among the 100-plus new buildings at the Bowman Army Air Base was the headquarters and administration building. This building held the office of the base commander and administrative staff of the air base. This may have been the first place that a newly assigned soldier to the base visited and the last one upon reassignment or discharge. (Edward Peck.)

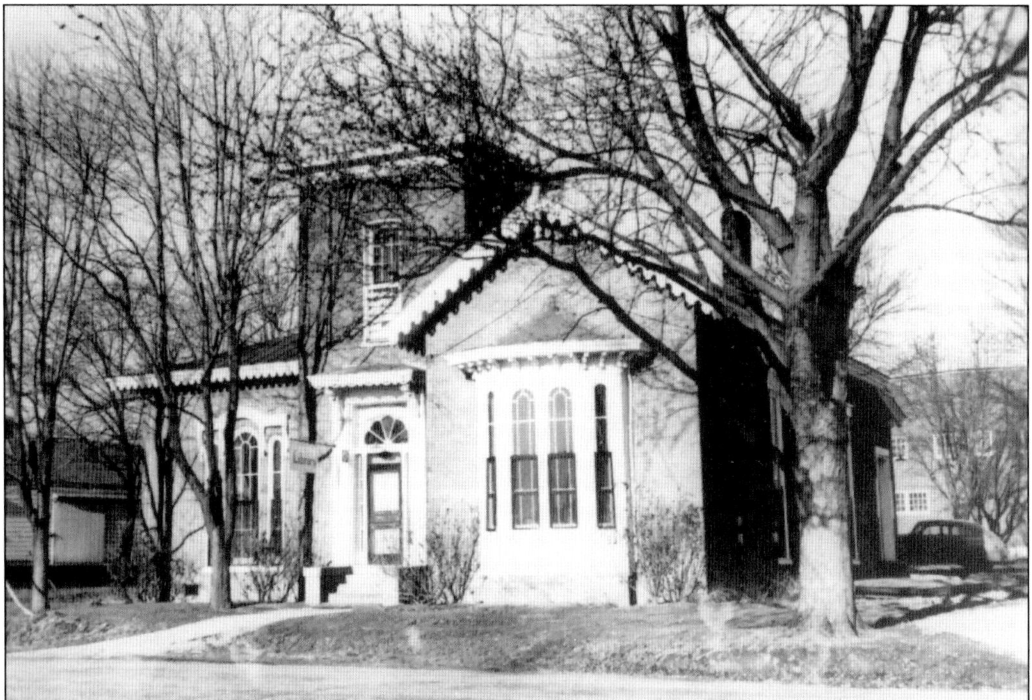

The Bowman Army Air Base library was situated in an old farm house that had long stood on the property. It was the only building on base to be surrounded by a stand of trees. Most interesting about the library was that it was staffed by members of the Crescent Hill Women's Club. They made this their wartime mission and featured its progress in the club newsletter. (Edward Peck.)

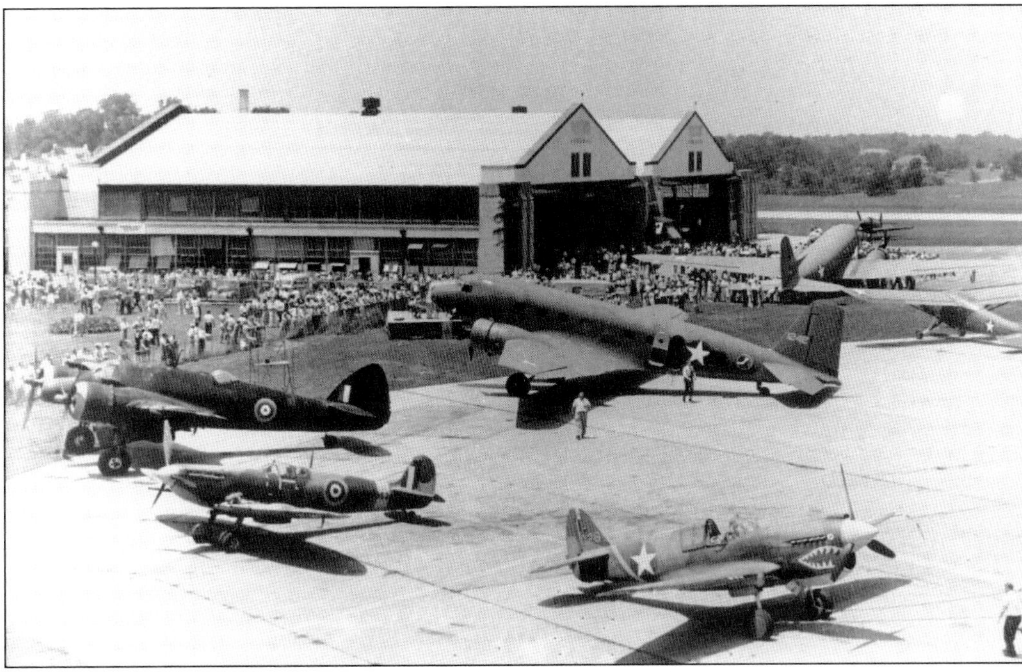

For the price of a 25¢ war stamp, Louisvillians were invited to the Bowman Army Air Base during the 1942 Fourth of July holiday to inspect a "galaxy" of airplanes during the War Bond Air Cavalcade Tour. The highlight of this tour was a flight of a captured German Me 109E fighter plane. Also displayed were a British Spitfire and Beaufighter along with an American P-39 and P-40. Over 30,000 visitors came during the three-day event, and over $3,500 was raised from selling war stamps. (Both, Edward Peck.)

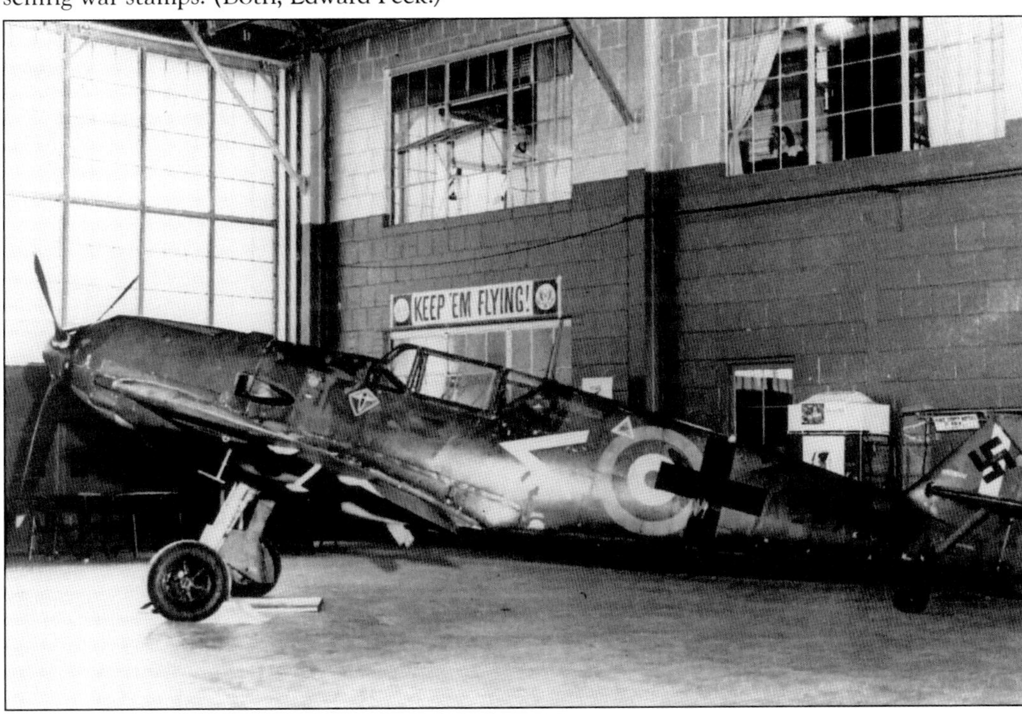

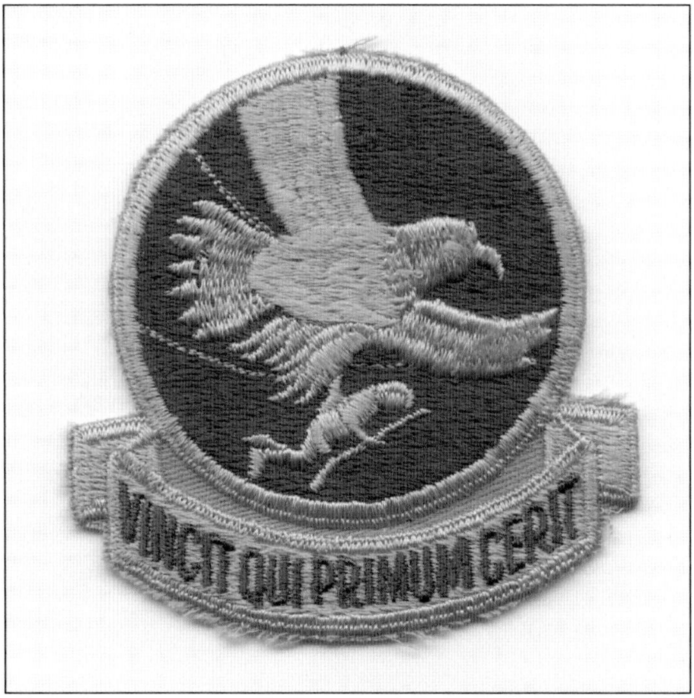

This was the insignia of the First Troop Carrier Command, often written as I Troop Carrier Command, during World War II. Its mission was to train troop carrier organizations and personnel. Its operating area was the United States, not overseas. First Troop Carrier was headquartered at Stout Field, Indianapolis. Its motto translates to "He conquers who gets there first." (Author's collection.)

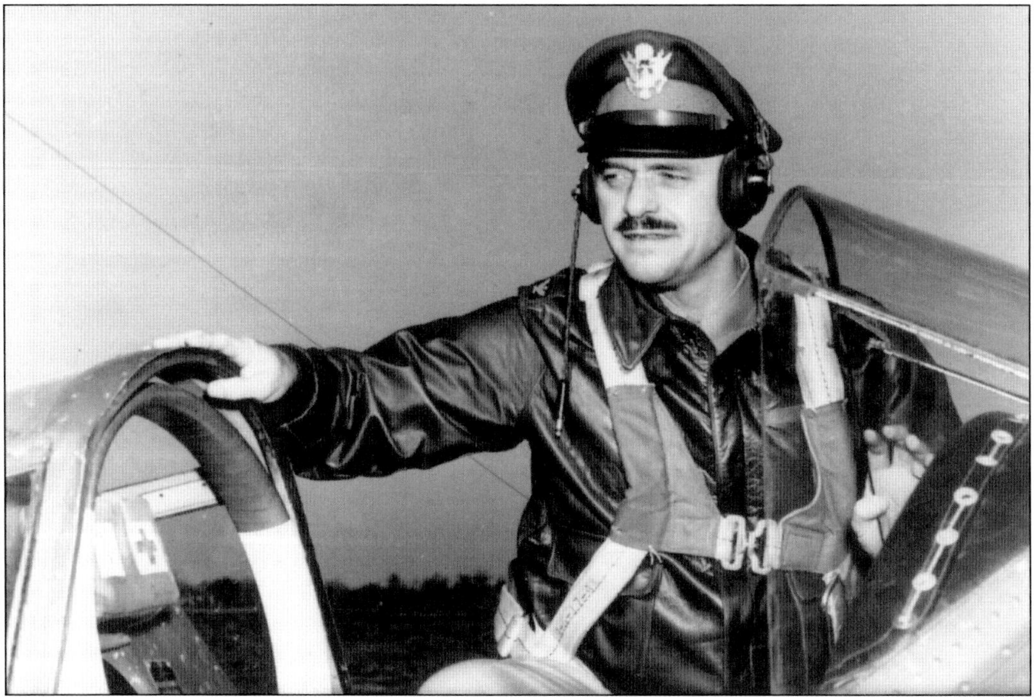

Col. George P. Johnson was the first base commander of the Bowman Army Air Base, taking command in December 1940. Although a native of New Jersey, his Army career started in 1917 and he came to Louisville from Langley Army Air Field, Virginia. He was reassigned to Stout Field in Indianapolis during 1945 but returned to Louisville after the war to spend the rest of his life there. He passed away in 1980. (Edward Peck.)

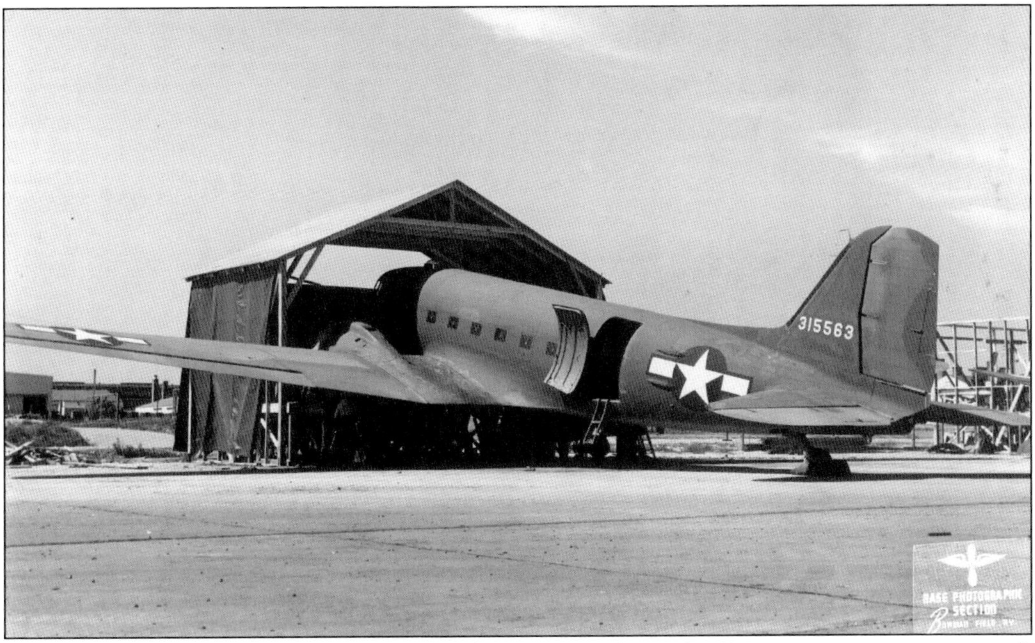

With the First Troop Carrier Command at the Bowman Army Air Base, troop transport aircraft became a common sight. Most common was the Douglas C-47 Skytrain. The C-47 was the military version of the DC-3 airliner; during World War II, the airplane became a legend for its ruggedness and dependability. At Bowman Field, C-47s carried troops and cargo, towed gliders, and helped train flight nurses. Since Bowman Field had few large hangars, C-47 maintenance was often done outside, and shelters known as "nose docks" were built to cover mechanics as they worked on engines. So important to the war effort was the C-47, Gen. Dwight D. Eisenhower wrote that its contributions were one of the reasons for winning World War II. (Above, author's collection; below, Edward Peck.)

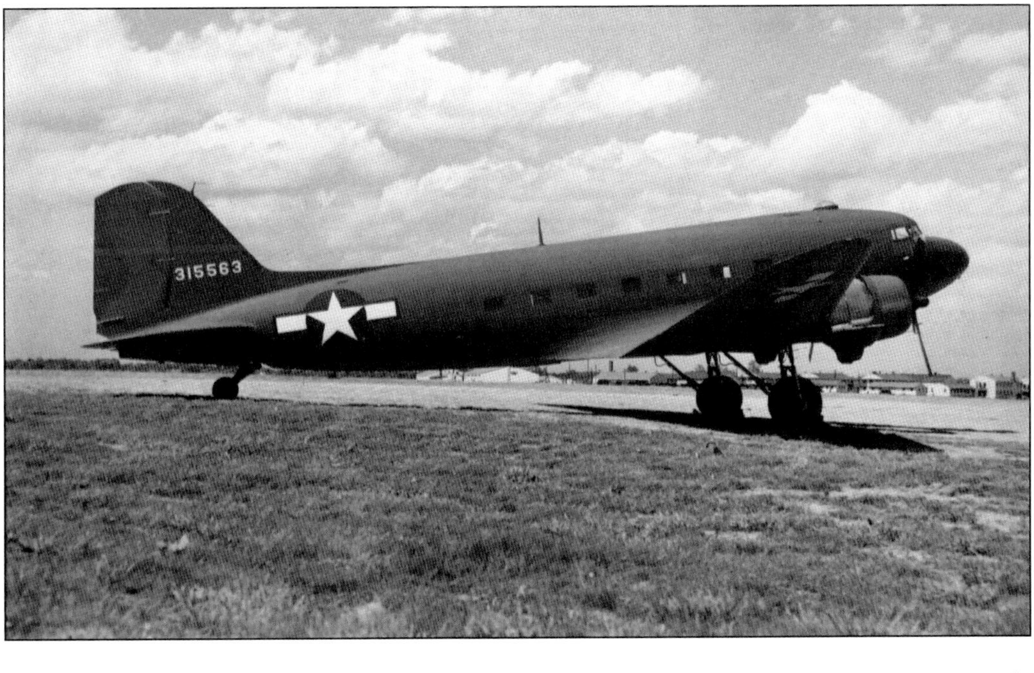

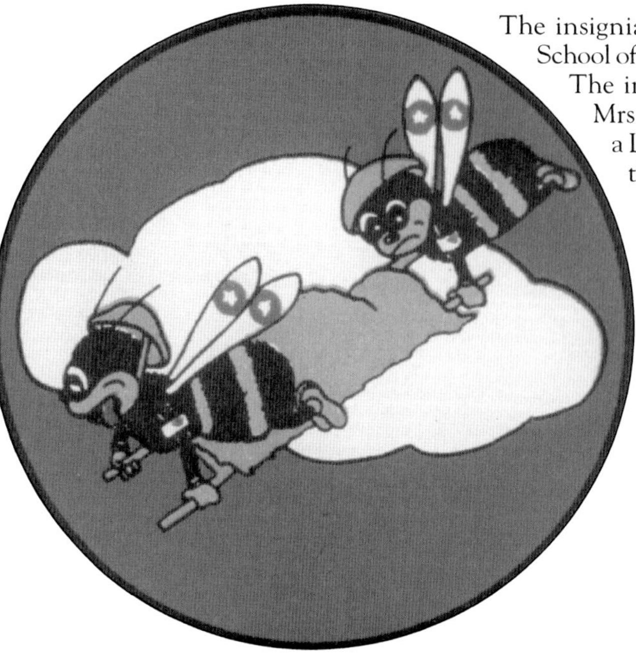

The insignia of the Bowman Army Air Base School of Air Evacuation featured honeybees. The insignia was said to be designed by Mrs. Don Rider of Buechel, Kentucky, a Louisville suburb. Variations such as this, with the bees carrying a stretcher in front of a cloud, are still issued to units in the current Air Force air evacuation community. (Wilma S. Ford.)

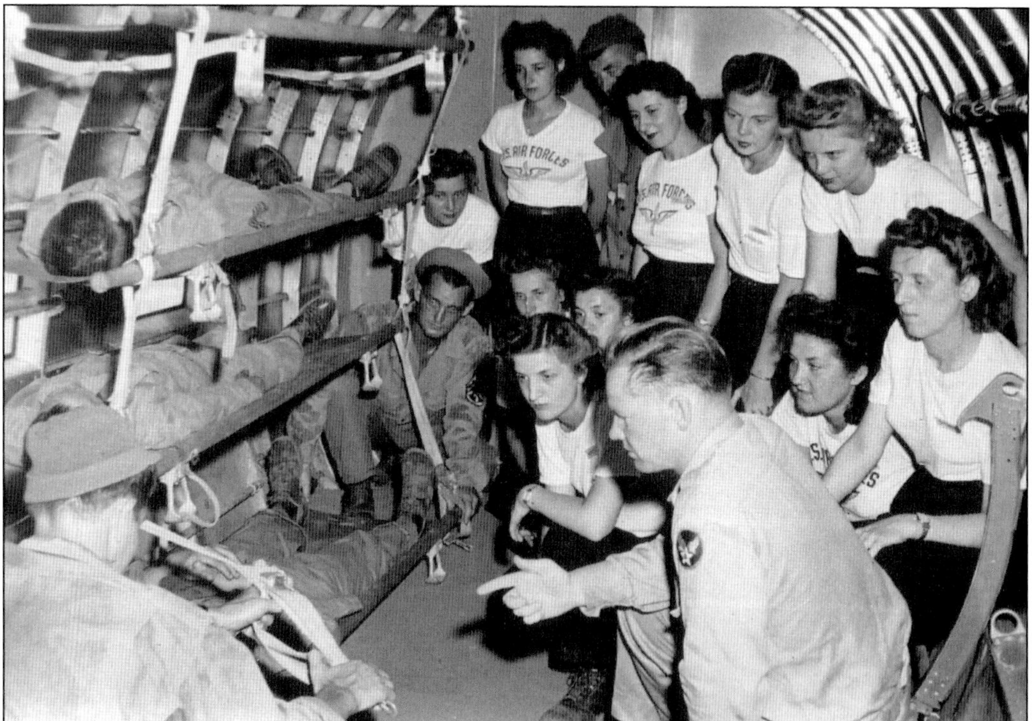

The Bowman Army Air Base School of Air Evacuation evolved over the two years that it was in Louisville. Arriving in October 1942, it was originally established as the 349th Air Evacuation Group. It was not until June 1943 that it took the school name. Classes were initially six weeks in length and then expanded to eight weeks. In this photograph, nurses are being instructed on how to care for patients inside an airplane. (Edward Peck.)

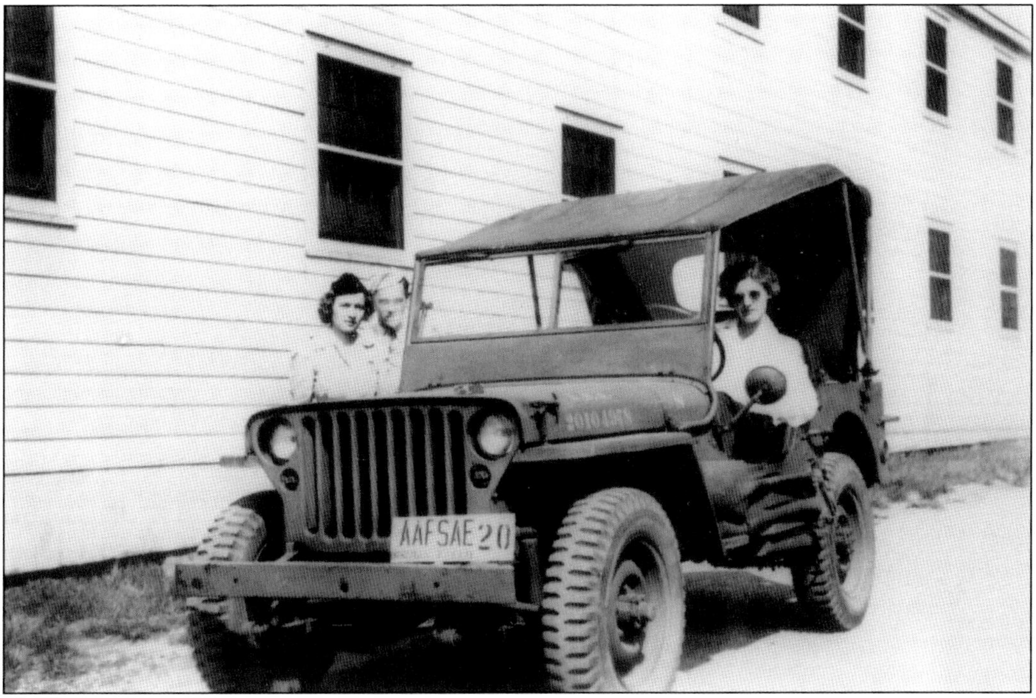

This jeep may look like thousands of others in use in the US military during World War II, but there is no mistake where this one came from—Army Air Forces School of Air Evacuation. Reading the plate over the bumper indicates that this jeep could only be found at the Bowman Army Air Base. Pictured below, ladies training at the school were all graduate nurses and were officers with at least the rank of lieutenant. (Both, Wilma S. Ford.)

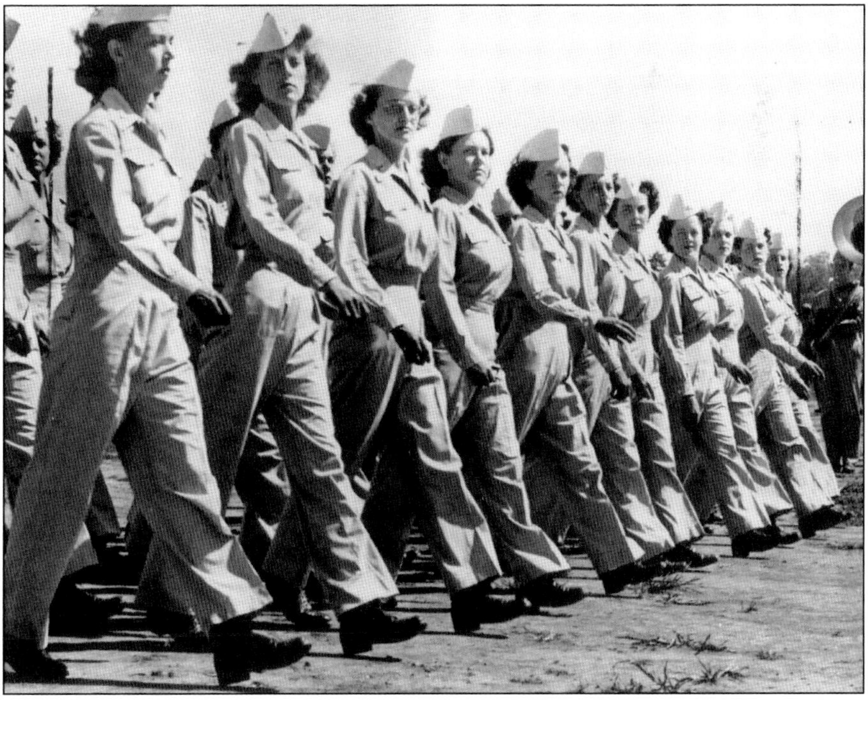

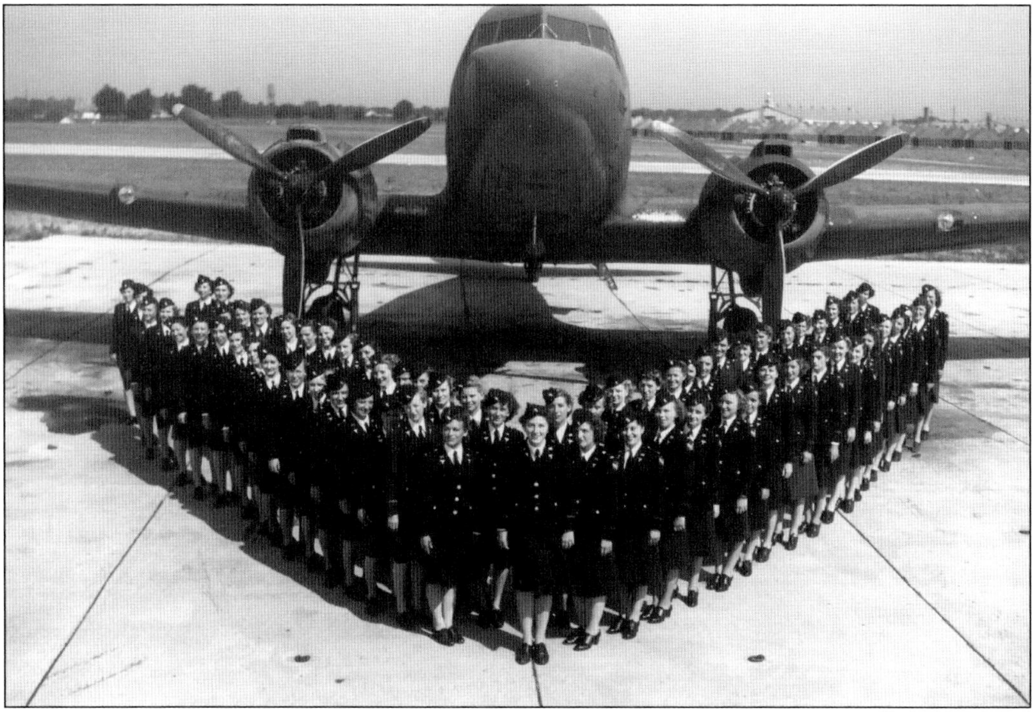

Flight nurses at the Bowman Army Air Base usually appeared in full dress uniforms for graduation or publicity photographs. These photographs probably indicate one of each. Above, nurses are wearing a dark blue uniform and have the insignia of the United States Army Air Forces on the shoulder. Since flight nurses were closely allied with troop carriers, a C-47 transport appears in the background. It was the C-47 that was the chief air evacuation transport during World War II. (Both, Edward Peck.)

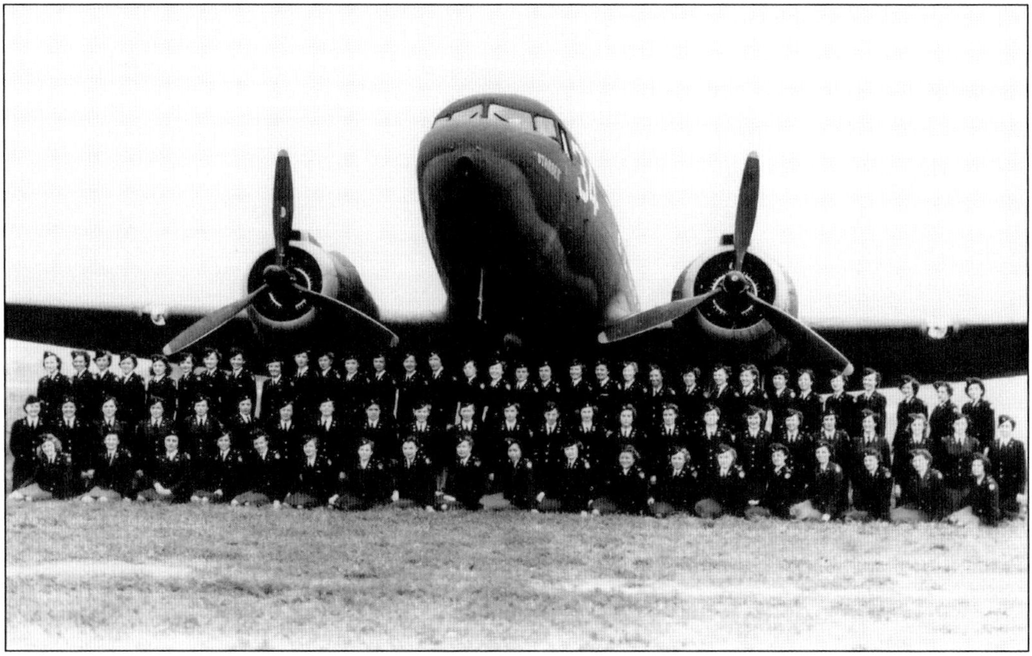

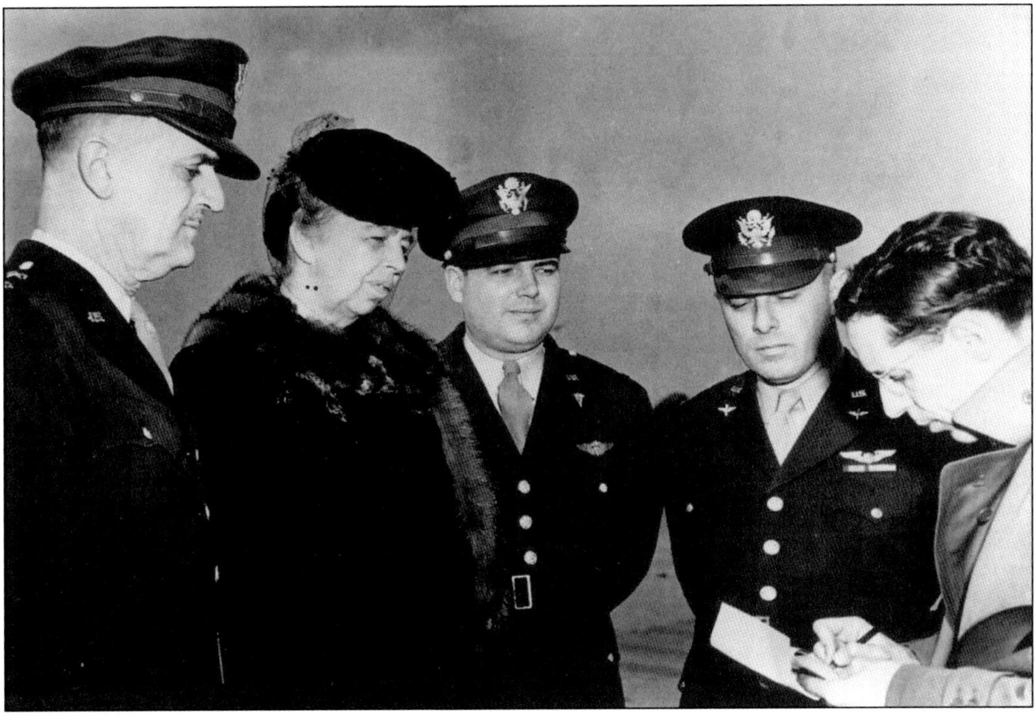

First Lady Eleanor Roosevelt is seen during her visit to the Bowman Army Air Base School of Air Evacuation on January 27, 1944. Impressed by a flight nurse demonstration that was performed for her, Roosevelt wrote, "These girls are the first who contact men after they are wounded and their work is essential." To her left is Maj. Gen. David N.W. Grant, surgeon general of the Army Air Forces. (Edward Peck.)

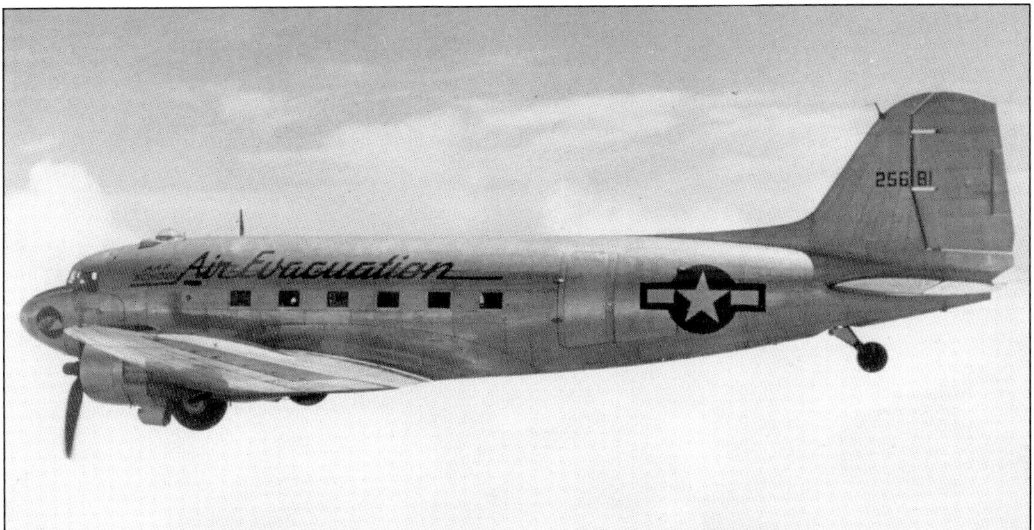

Flying high, this C-47 transport was assigned to the School of Air Evacuation at Bowman Field. It was used so flight nurses could practice their skills in the air where the environment was noisy and constantly moving. Other uses of this C-47 included visits to other bases for flight nurse demonstrations, and appearances at air shows. (Chester Lamppin.)

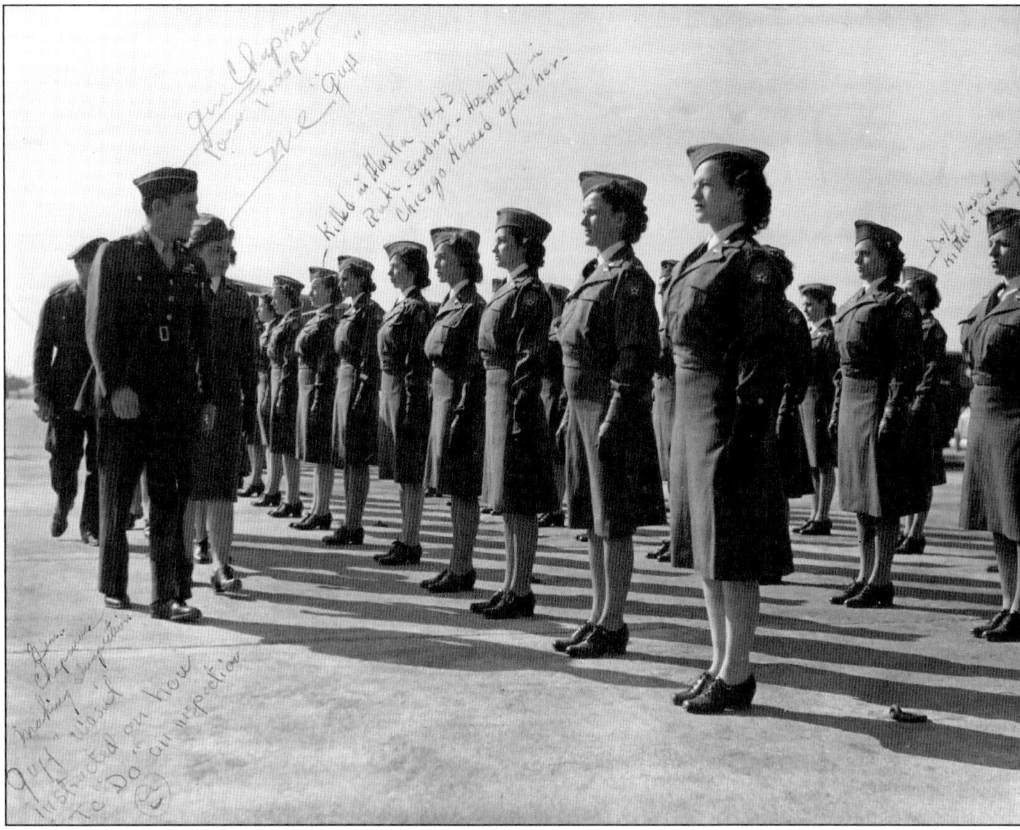

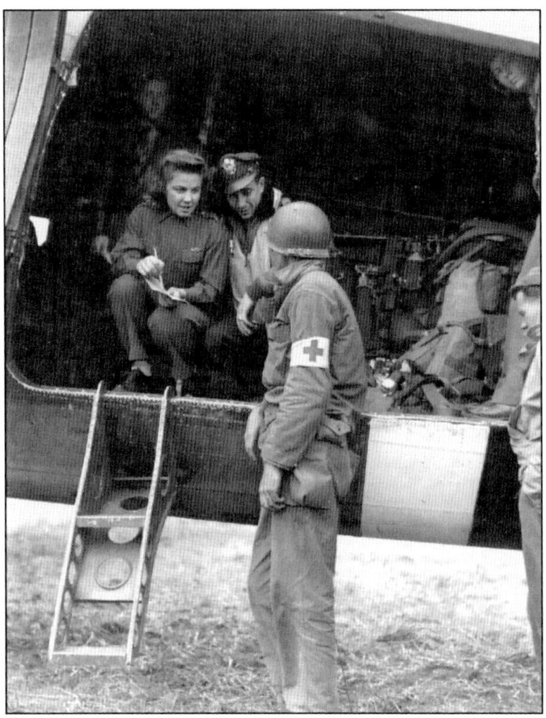

The first "official" graduate of the Bowman Army Air Base School of Air Evacuation was Lt. Ethel Guffey from Shawnee, Oklahoma. Undoubtedly, it was her leadership qualities and outstanding personality that earned her this honor during her graduation on February 18, 1943. Her wartime service as a flight nurse included Normandy, Holland, and Germany. She is seen at left inside a C-47 preparing for an air evacuation mission. (Both, Ethel Guffey Simpson.)

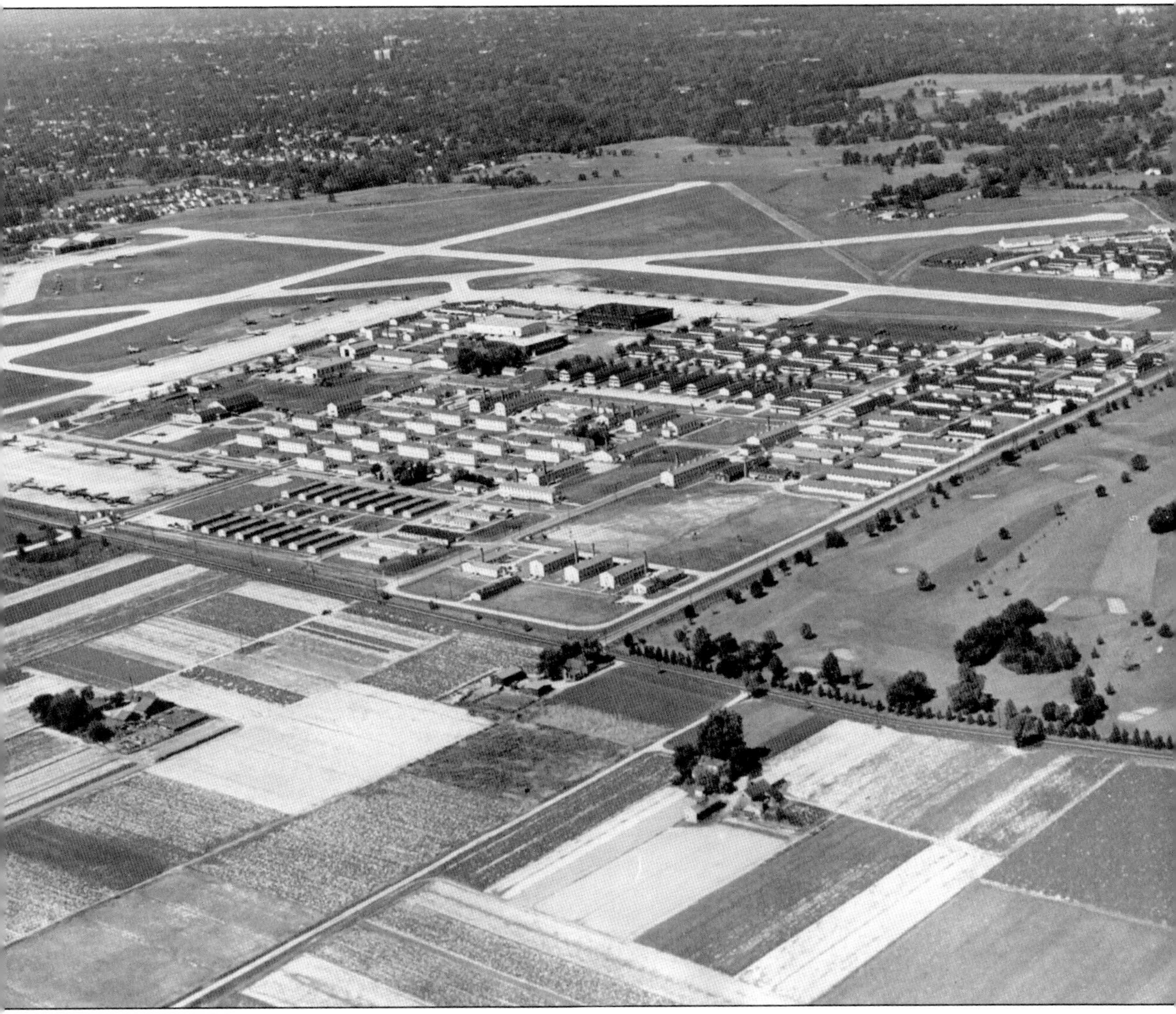

This aerial view shows the progress of building construction at the Bowman Army Air Base. When the last nail was driven, 124 new buildings were completed. Construction started in 1940 and was completed in 1944, with additions for the School of Air Evacuation. After the war, demolition would take years, with the last building removed during the 1980s. Only the big east-side hangar remains from wartime construction, although traces of buildings can still be found. (Courtesy of Edward Peck.)

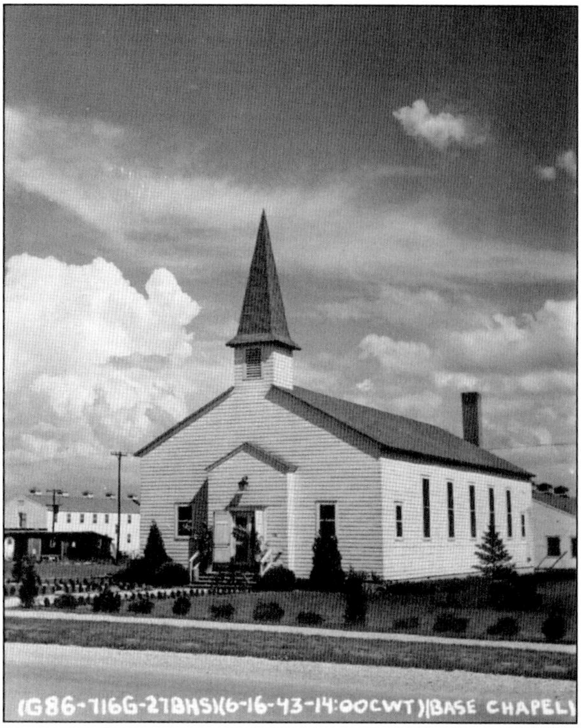

The Bowman Army Air Base chapel had its first Sunday service conducted on August 30, 1942, with chaplain Capt. James R. Crowe (below) in charge. This chapel was the home for all faiths and was the location of a number of wartime weddings among base personnel. After the war, the chapel was moved about two miles to Browns Lane and became the sanctuary for St. Matthews Methodist Church. It was demolished in 1989 to make room for a new building. (Both, Edward Peck.)

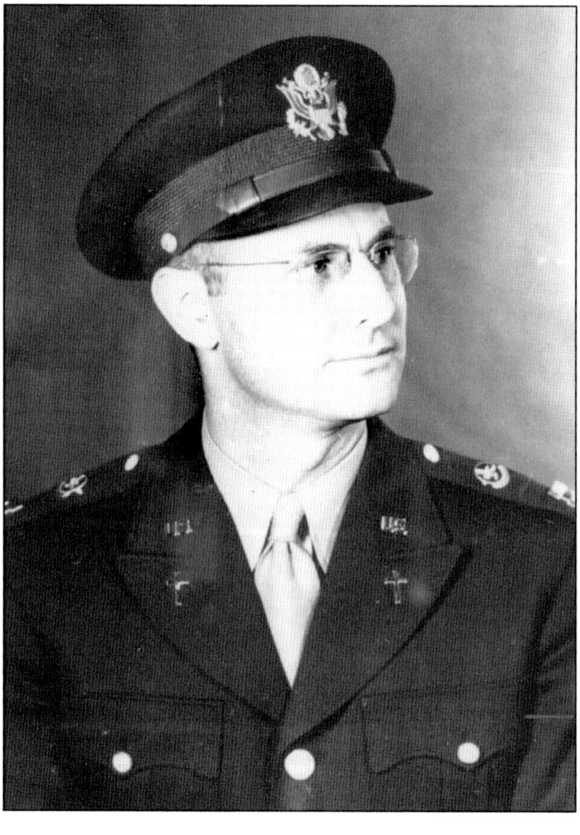

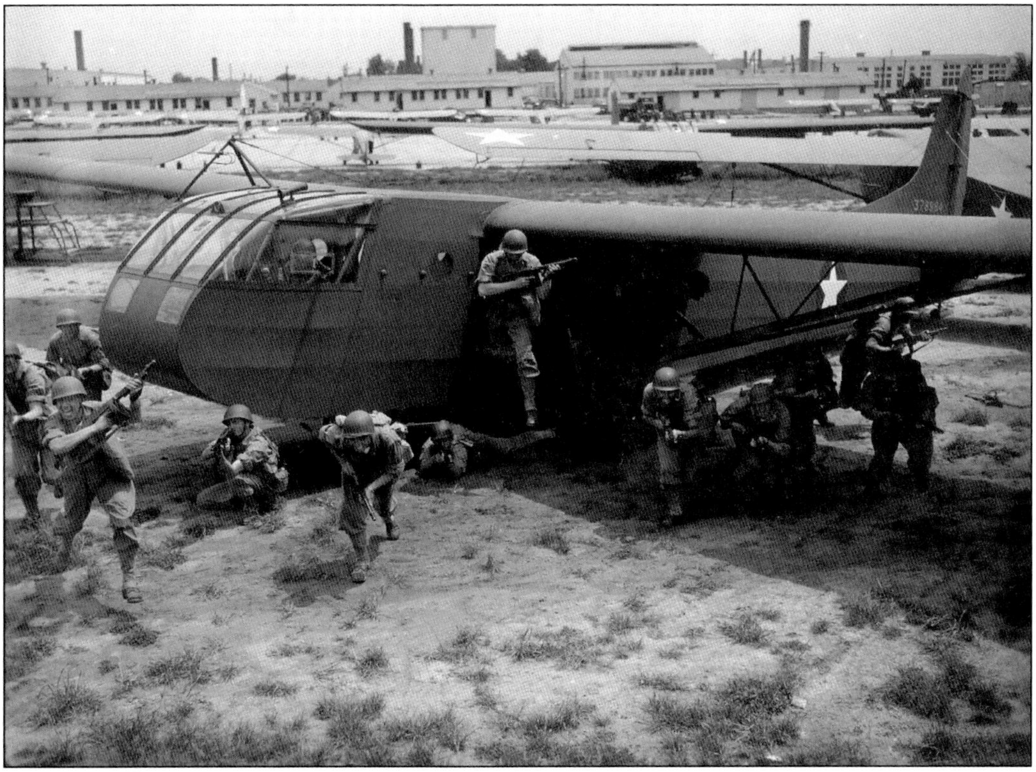

Since the mission of the Bowman Army Air Base was training troop carrier aircraft crews, the addition of the Glider Combat Training Center in 1943 was logical. At Bowman Field, training was conducted on the WACO CG-4A glider nicknamed the "Hadrian." It could carry 13 men or a jeep or a small artillery piece or cargo. The nose section could hinge upward for convenient loading and unloading. Above, soldiers demonstrate a quick exit from a CG-4A, while below, another glides to a landing. (Both, Edward Peck.)

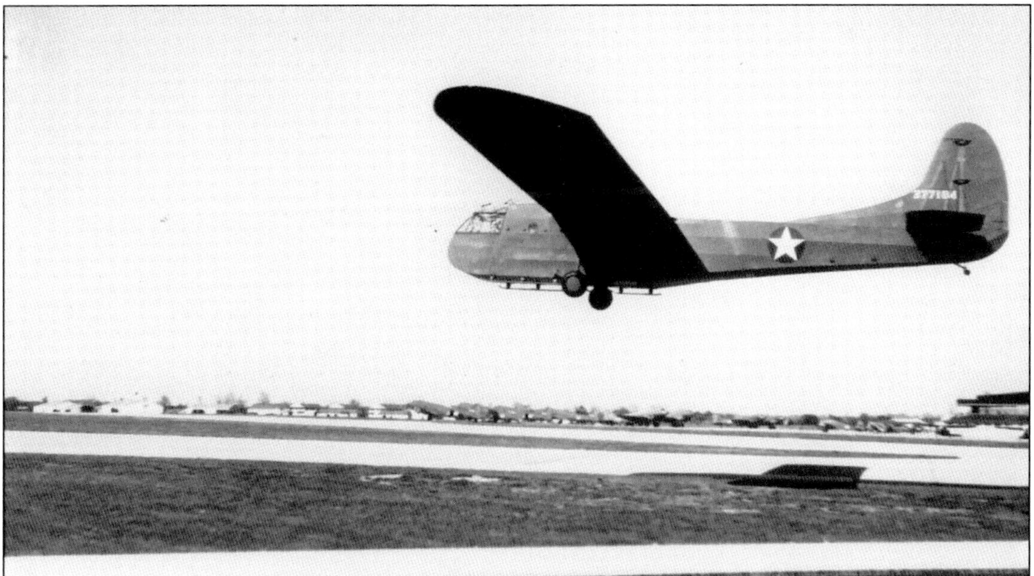

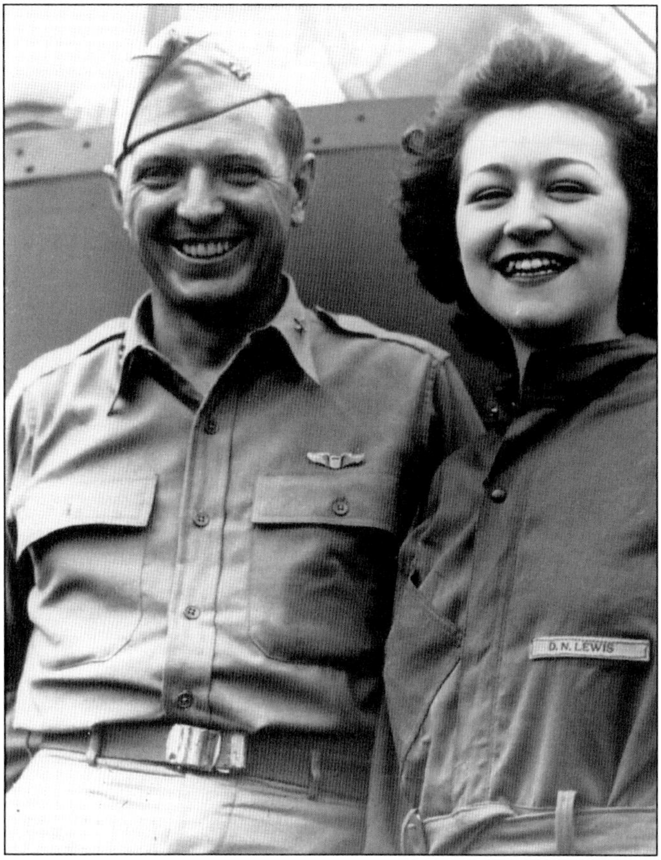

Maj. Mike Murphy, a well-known air show pilot before the war, was the chief glider instructor pilot with the First Troop Carrier Command. He was assigned to the Bowman Army Air Base through much of 1943. He was the primary spokesman for the glider program at Bowman Field, often flying demonstrations or giving glider rides to the press or local dignitaries. He was assigned to fly the first glider into Normandy during the D-Day invasion on June 6, 1944. (Edward Peck.)

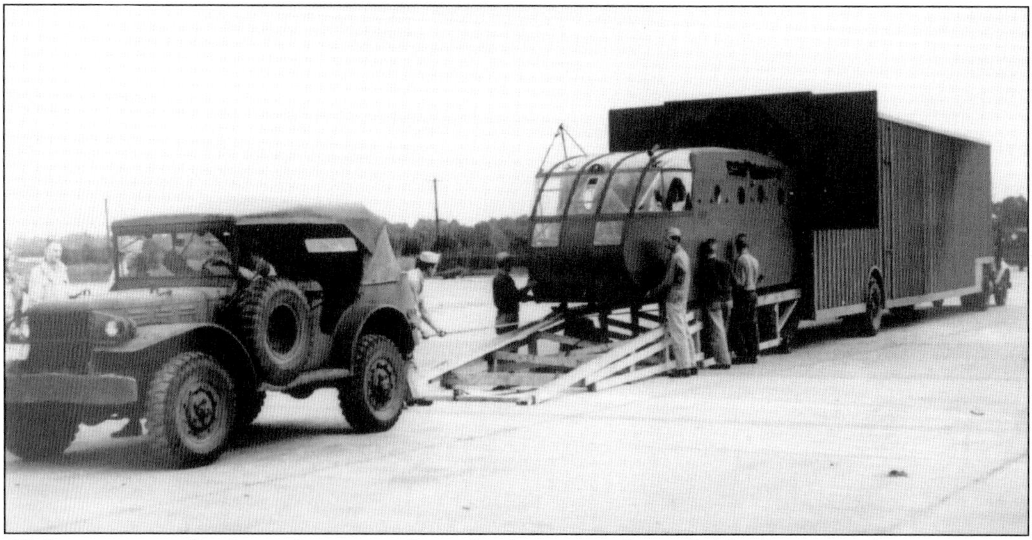

Gliders were usually shipped in crates from the factory to the airfield. They were then assembled and flight tested. Nearly 14,000 WACO CG-4A gliders were made during World War II and not all by the aircraft industry. Some were made by the Ford Motor Company and Gibson Refrigerator Company. Gliders at Bowman Field numbered in the dozens. The majority were saved for combat. (Edward Peck.)

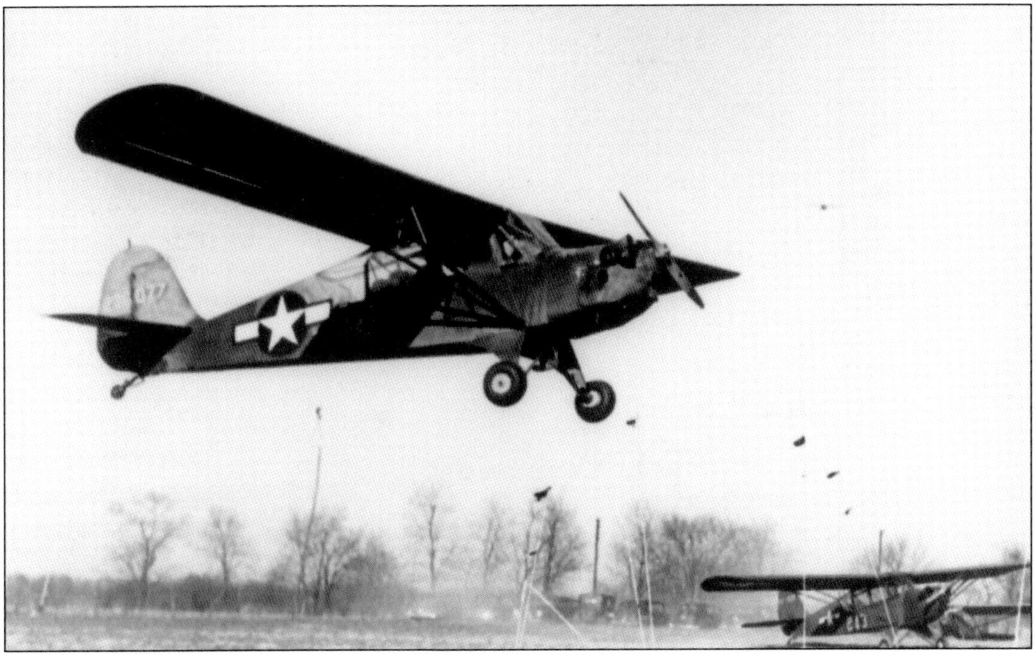

Since only a small number of gliders were available at the Bowman Army Air Base for training, glider pilots would maintain their flight proficiency in small Grasshopper airplanes, mainly Taylorcraft L-2s and Aeronca L-3s. One training assignment called for the pilot to land his airplane in a field "dead stick," or with the motor turned off, to simulate a glider landing. This could be dangerous, and some fatalities occurred from crashes. (Both, Edward Peck.)

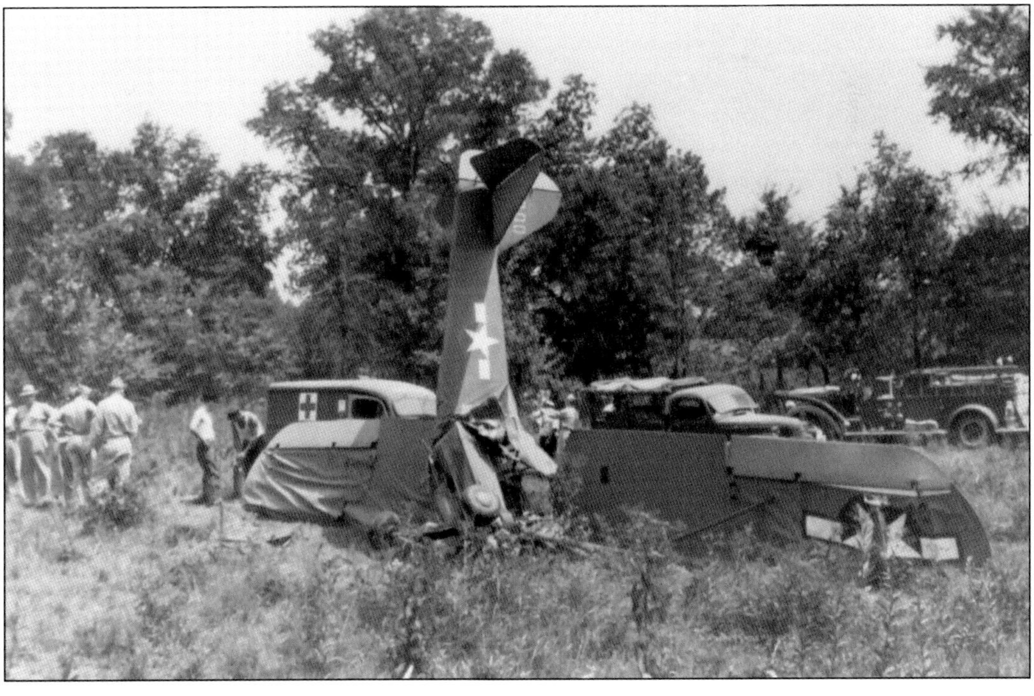

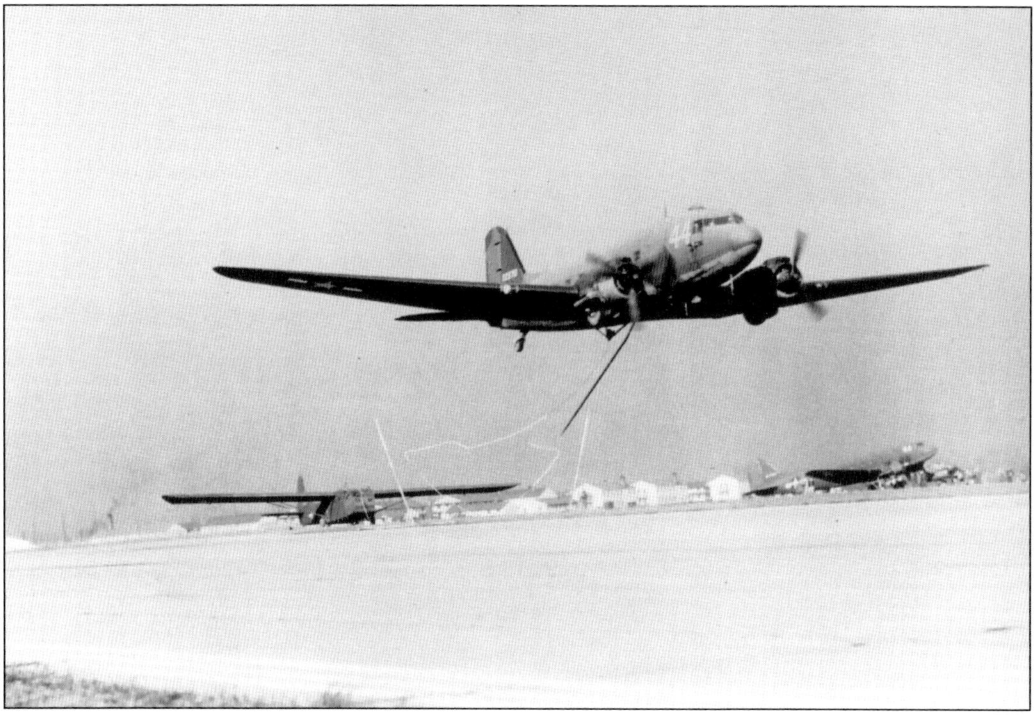

Although the combat glider was considered expendable in war, some could actually be reused if they could be removed from the combat zone. To accomplish a recovery, a procedure called the "glider snatch" was devised, in which a low-flying transport airplane would snatch a glider into the air. This was practiced at the Bowman Army Air Base. One benefit of a recovery was that the glider could be loaded with wounded soldiers for transport to a hospital. (Both, Edward Peck.)

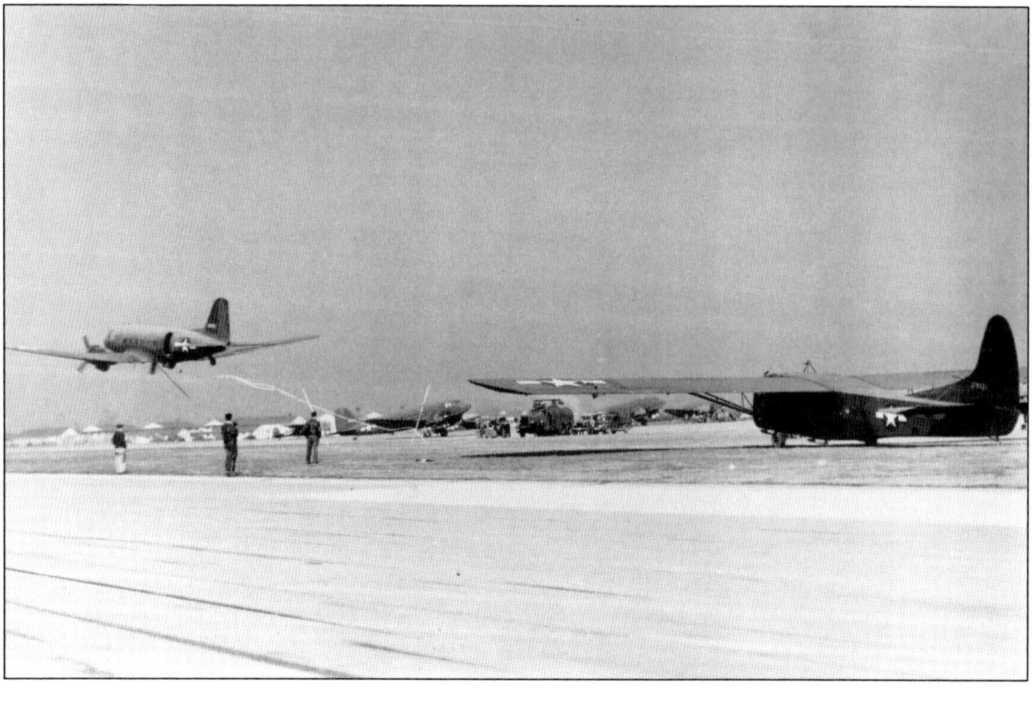

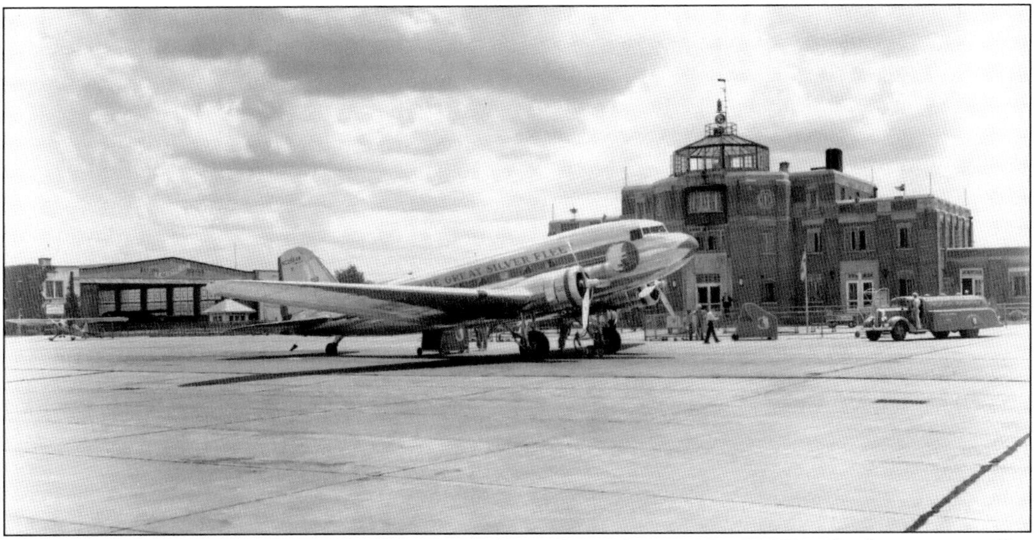

While Bowman Field was converted to a military base during World War II, there was still a civilian presence at the field. Both American and Eastern Airlines continued daily flights in and out of Louisville. Moreover, Louisville Flying Service continued to serve civilian aviation. This Eastern Air Lines DC-3 stands by for its next flight on what looks like a most pleasant day. (Edward Peck.)

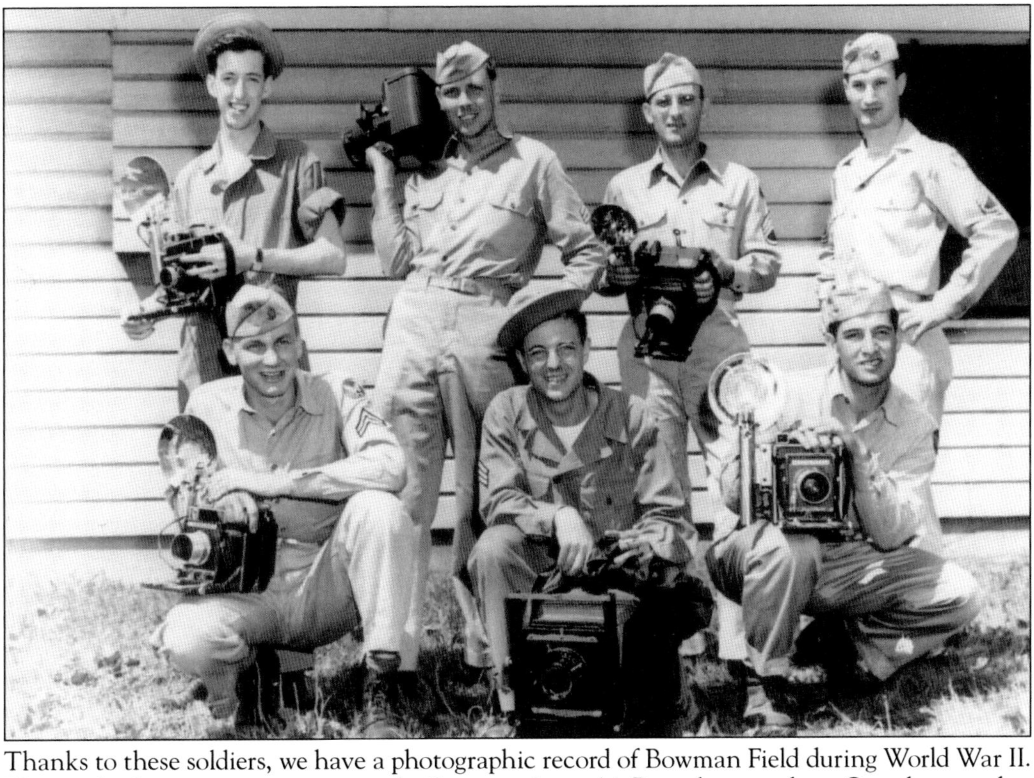

Thanks to these soldiers, we have a photographic record of Bowman Field during World War II. Waiting for their next assignment are the Bowman Army Air Base photographers. One photographer assigned to the photography shop was Ray Murphy, who, after the war, opened the Murphy's Camera Center chain, which still serves Louisville and surrounding cities. (Edward Peck.)

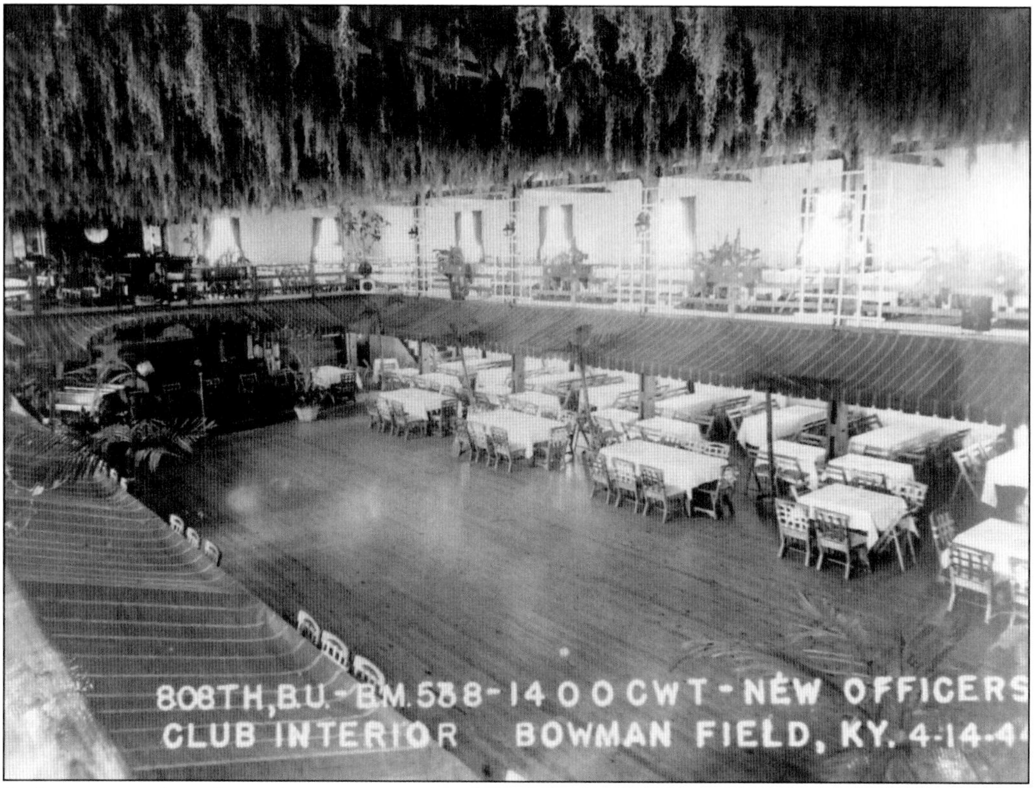

In keeping with military bases across the country, an officers' club was an absolute essential. Above, the "Hanging Gardens" of the Bowman Army Air Base Officers Club are on full display. Membership had to be approved, and a membership card was issued to each applying officer. Just as important was the need for a service club where personnel could relax, enjoy a beverage, and listen to music from a jukebox. (Both, Edward Peck.)

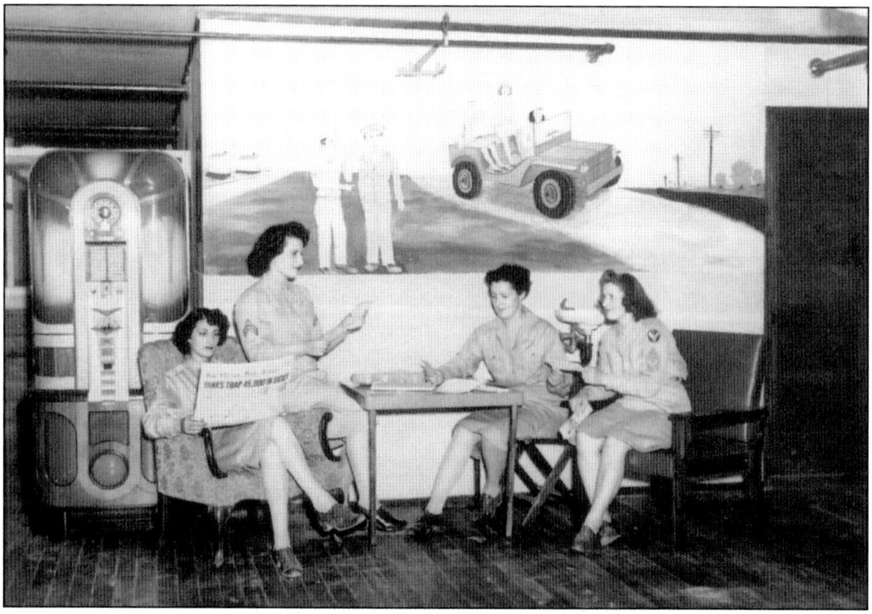

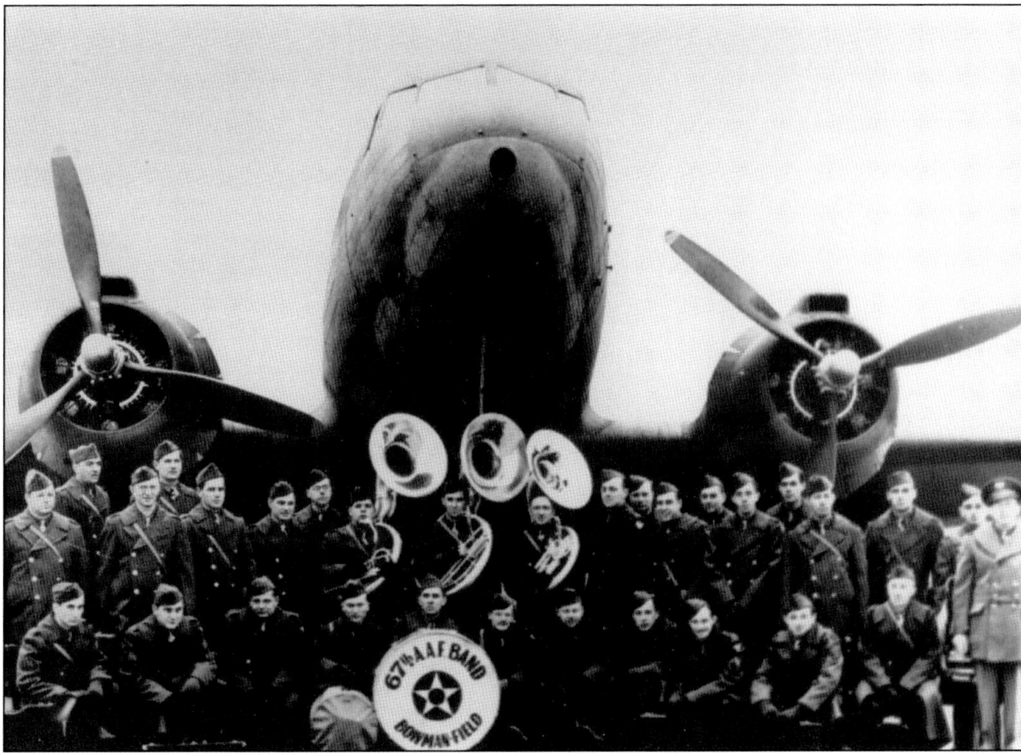

One of the busiest units at the Bowman Army Air Base was the 67th, later redesignated 567th, Army Air Forces Band. Led by CWO Wessel Peonie, the band not only performed for the air base but also went out into the community to perform for local businesses that made products for the war effort. (Edward Peck.)

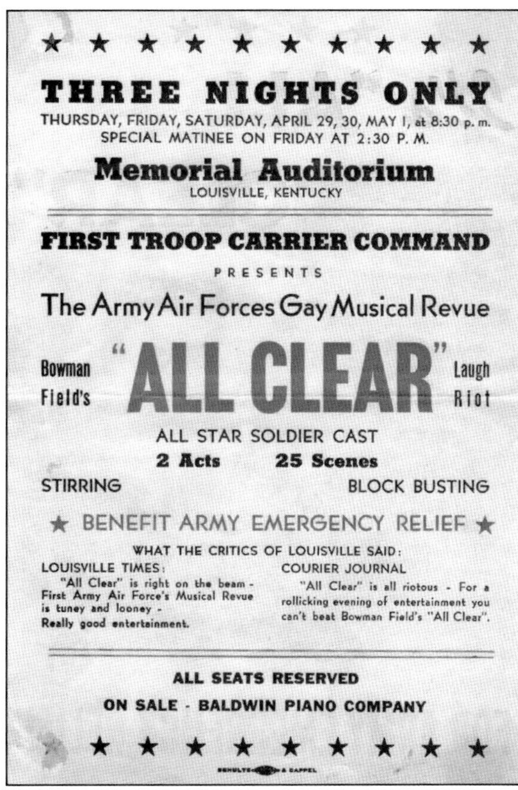

One of the most successful ventures at the Bowman Army Air Base was the creation of a very professional musical revue titled *All Clear*. With 14 songs, *All Clear* depicted life in the Army Air Forces. It premiered at Louisville's Memorial Auditorium on February 11, 1943. A record album was released, with "Da Da Dit" the hit song. It was about a soldier telegraphing his love to his sweetheart. A songbook was also published. The revue went on the road to nearby cities and then made a return engagement in Louisville. (Author's collection.)

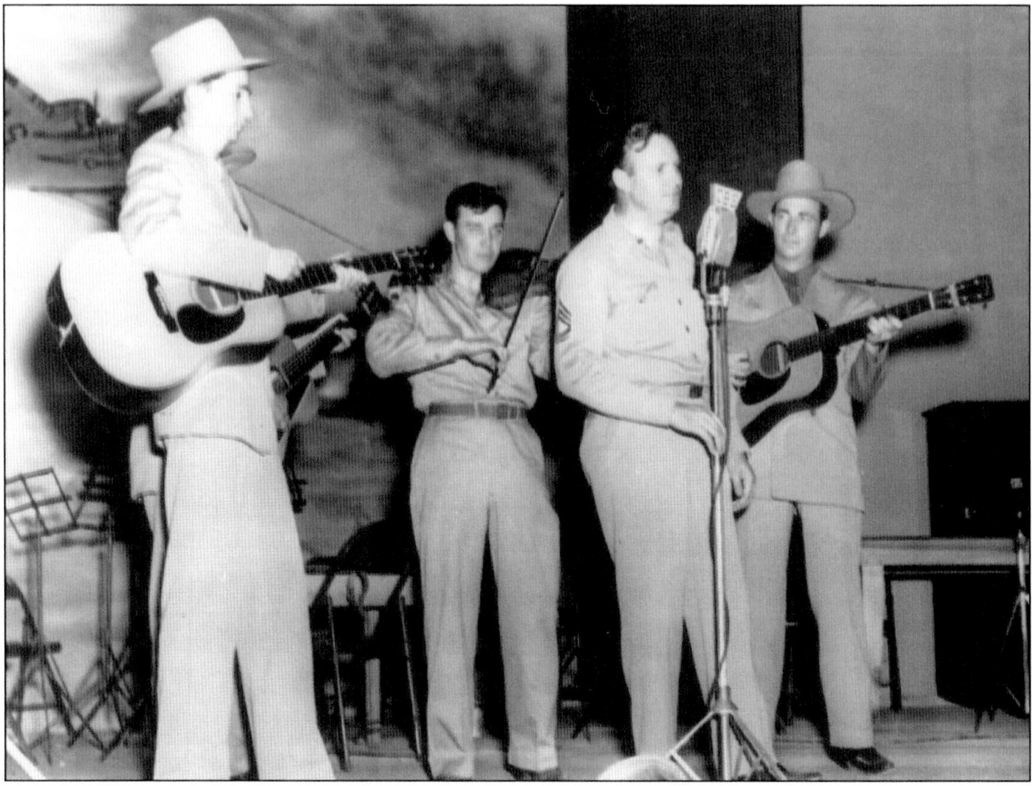

Cowboy singer Gene Autry was at the Bowman Army Air Base in June 1943 to broadcast his popular radio program from the airfield. In the photograph of Autry and his band above, he might be singing "Back home in Louisville," if there was such a song, because, in 1933, he lived here and worked for WHAS Radio. Autry arrived in a B-17 named "Mary Ann: Star of Air Force." This is in reference to its flying role in the movie *Air Force*, which was filmed right after the attack on Pearl Harbor. (Both, Edward Peck.)

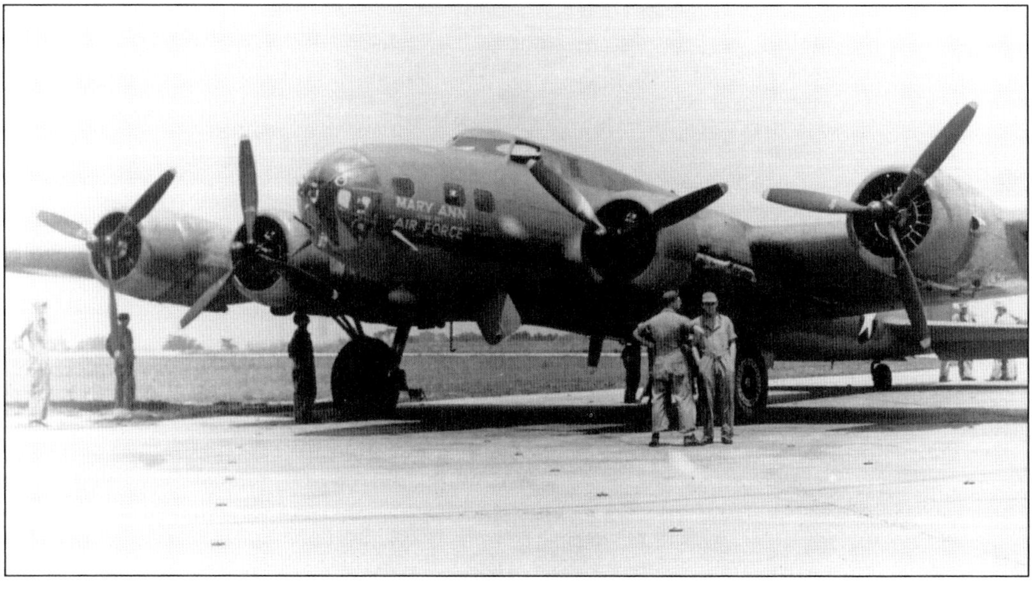

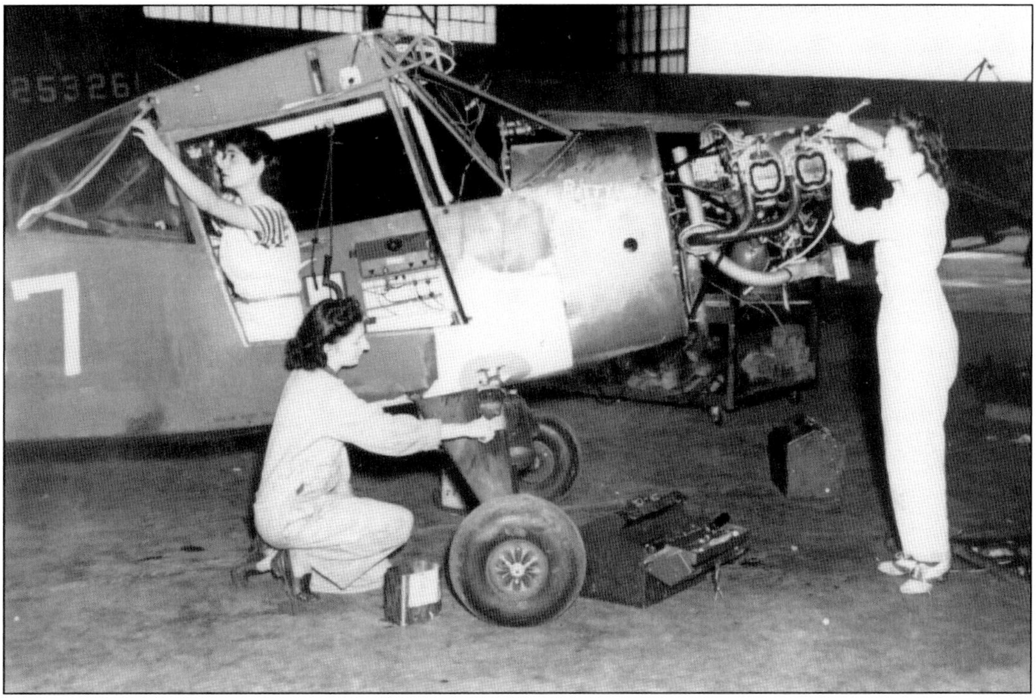

With so many men in uniform during World War II, women were needed to take on many tasks. At the Bowman Army Air Base, civilian women filled many positions from managing the snack bar to aircraft maintenance. In this photograph, these maintainece workers do inspection and repair on one of the many Grasshopper airplanes flown by glider pilots. (Edward Peck.)

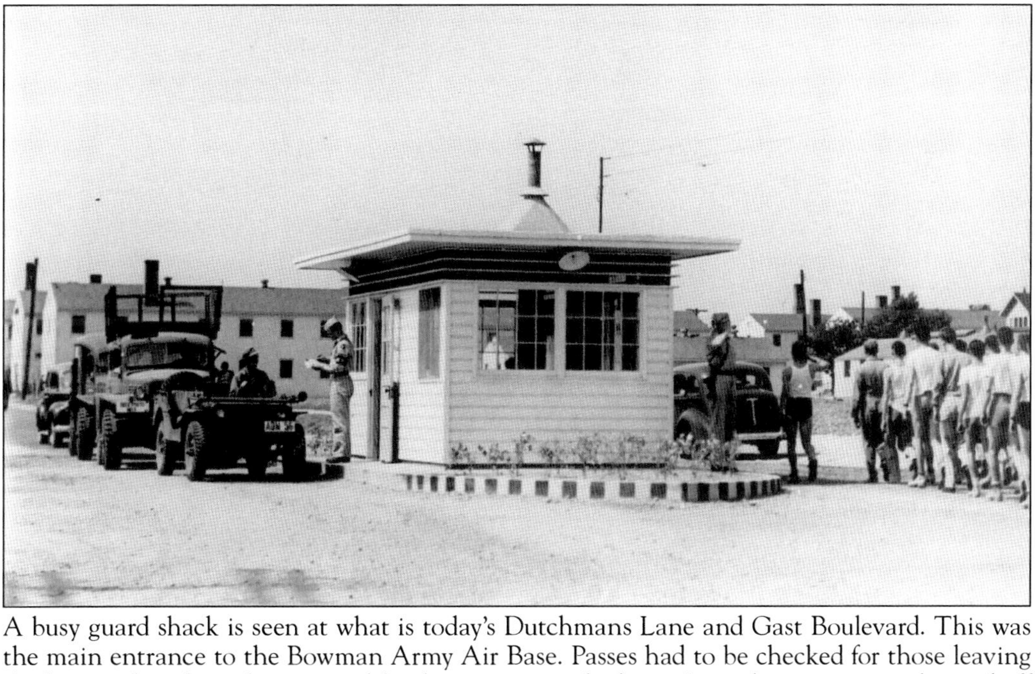

A busy guard shack is seen at what is today's Dutchmans Lane and Gast Boulevard. This was the main entrance to the Bowman Army Air Base. Passes had to be checked for those leaving the base and credentials inspected for those entering the base. A north entrance was located off today's Cannons Lane. No evidence of this can be seen today at Bowman Field. (Edward Peck.)

The BOWMAN

BUZZ

PUBLISHED EACH SATURDAY
AT BOWMAN FIELD

VOL. II NO. 35 AN AAF PERSONNEL DISTRIBUTION COMMAND STATION 29 SEPTEMBER

FIRST BOWMAN DISCHARGEE

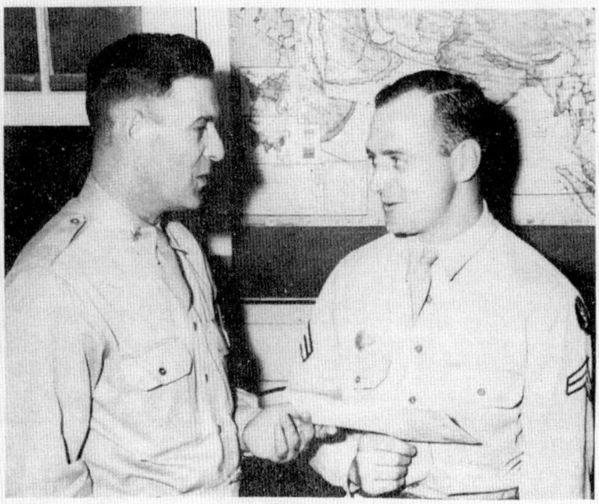

First tangible result of Bowman Field's Separation Point is pictured above as Corporal Jesse F. Boyer, Duncannon, Pa., receives his discharge papers from Major Charles T. Munn, commanding officer of Squadron A. Under the processing schedule followed here it requires less than 36 hours for complete processing and discharge of an individual.

BASE UNITS ARE CONSOLIDATE
TWO NCO CLUBS WILL MERGE

OFFICERS COMBINE THEIR CLUBS, NEW BOARD ELECTED

The proposed consolidation of the Bowman Field and PDC NCO Clubs will take place at an early date, it was announced this week by T/Sgt. Charles Whalen, president of the Bowman Field club, and T/Sgt. Ralph Boschert, president of the PDC club.

Effective at once will be the moving of the 1077th club to quarters on the base. The new cite will be the old Red Cross Snack Bar in Building 441. The move is expected to be completed this week.

When the consolidation of the two organizations is final, a new board of officers, half from each club, will be elected. Both clubs, the one at building 441, and the one downtown at Fifth and Market will be maintained. Membership in

NON-COMMISSION PERSONNEL OF P SHIFTED TO 1077

A sweeping reorganization both the 1000th and 1077th I Units, which in effect w consolidate most of the He quarters unit with the Base ganization, was announced ficially this week by Ma General Hubert Harmon. consolidation move will be fective October 1.

In substance, the order reorganize will transfer enlisted personnel now attac to the 1000th B.U. to 1077th. It will also trans to the 1077th all PDC offic who are not assigned to d with staff offices of Headq ters, and whose chief funct will be the administration the Headquarters squadron the Base.

1077th REORGANIZES SQDNS

The Bowman Buzz was the newspaper issued by the Personnel Distribution Command (PDC) when it was at the Bowman Army Air Base. After November 1, 1944, when all training at Bowman Field ended, the base was converted to a convalescent center for returning soldiers. This center helped ease soldiers back into civilian life after months of both physical and mental stress from battle. At the same time, the Army Air Forces headquarters for the PDC moved to downtown Louisville. (Becky Carnes.)

Four

Bowman Field Takes Flight

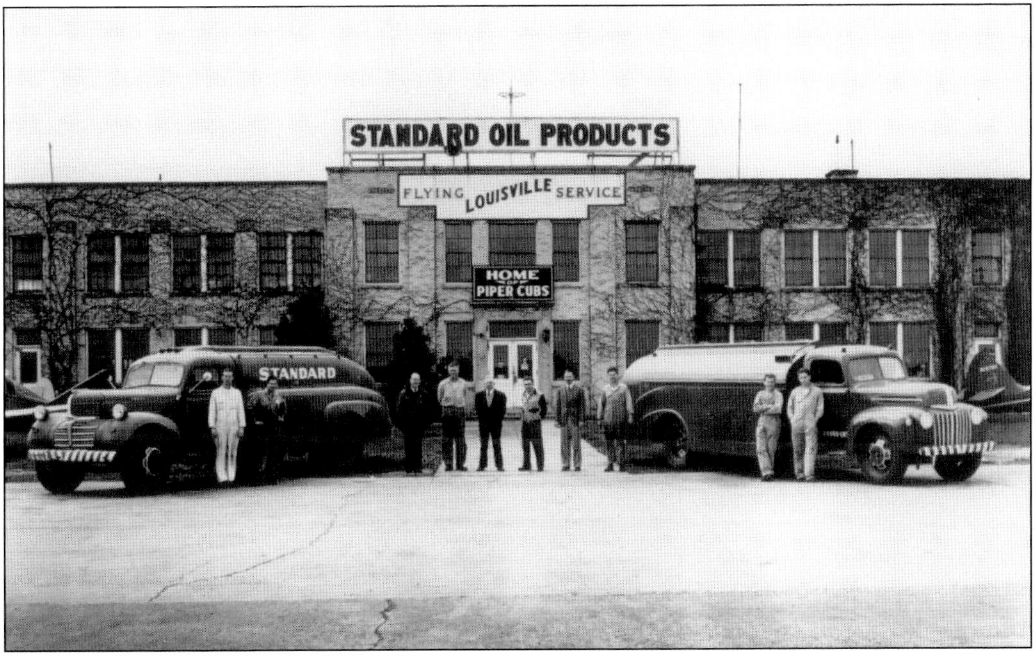

With the end of World War II in 1945, Bowman Field slowly transitioned from a military base back to a civilian airport. Louisville Flying Service was still in the old Curtiss Flying Service hangar and ready to serve the postwar needs of Louisville aviation. Anticipating a major increase in business, it moved to the twin hangars opposite the administration building in 1950. Standing between the trucks in this late 1940s photograph are John C. Bennett (third from left) and Frank Holzaphel (second from left). Louisville Flying Service employees are also pictured. Of interest is the ivy-covered north face of the building. (Edward Peck.)

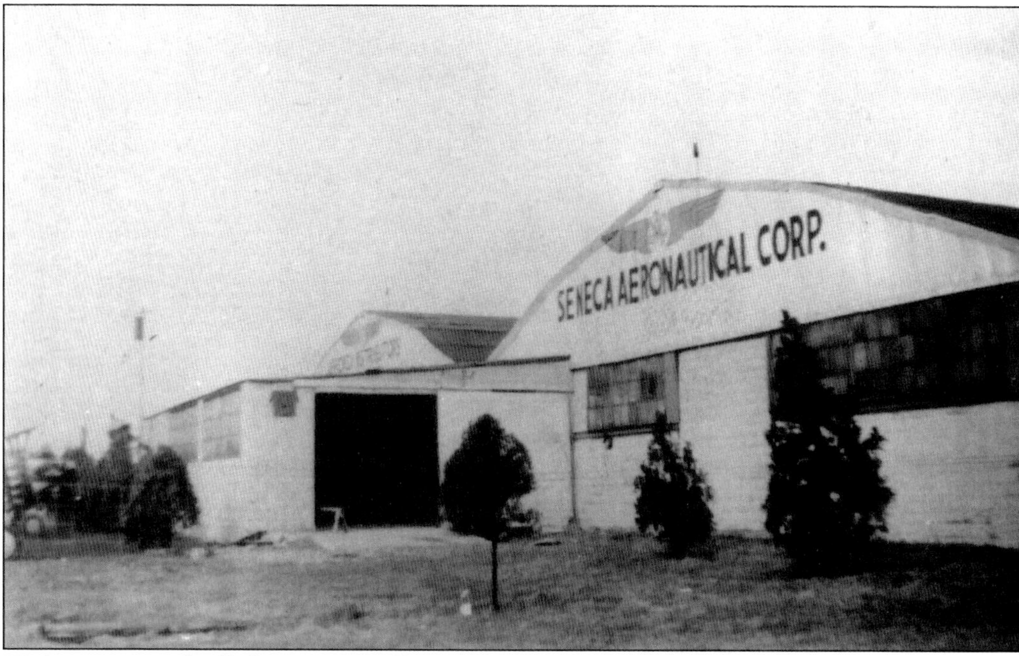

With provisions for flight training offered in the GI Bill for returning veterans, schools to train pilots popped up virtually overnight at Bowman Field. In all, 22 were established and located across the airfield. Most were short-lived because of lack of demand or the benefits of the GI Bill ending. One was Seneca Aeronautical, which was located in the original metal hangars dating to the 1920s and repositioned to the east side of the airport. It was previously occupied by Fall Cities Flying Service and after by Wade Flying Service. Robinson Flying Service was another one and was located in one of the wartime buildings. Next to it is Central American Airways, which still exists at Bowman Field although in another facility. (Both, author's collection.)

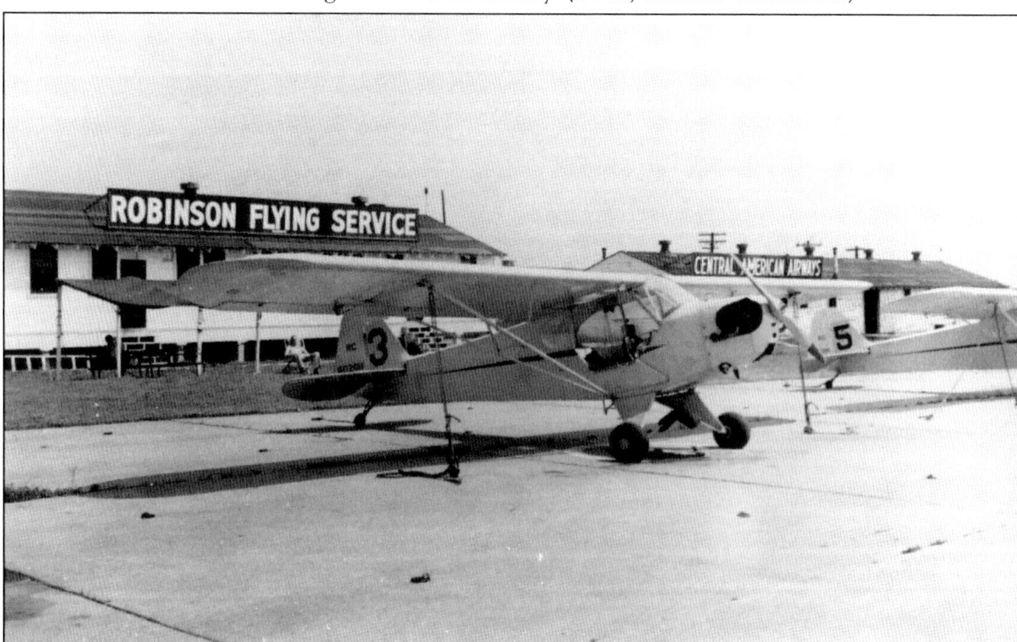

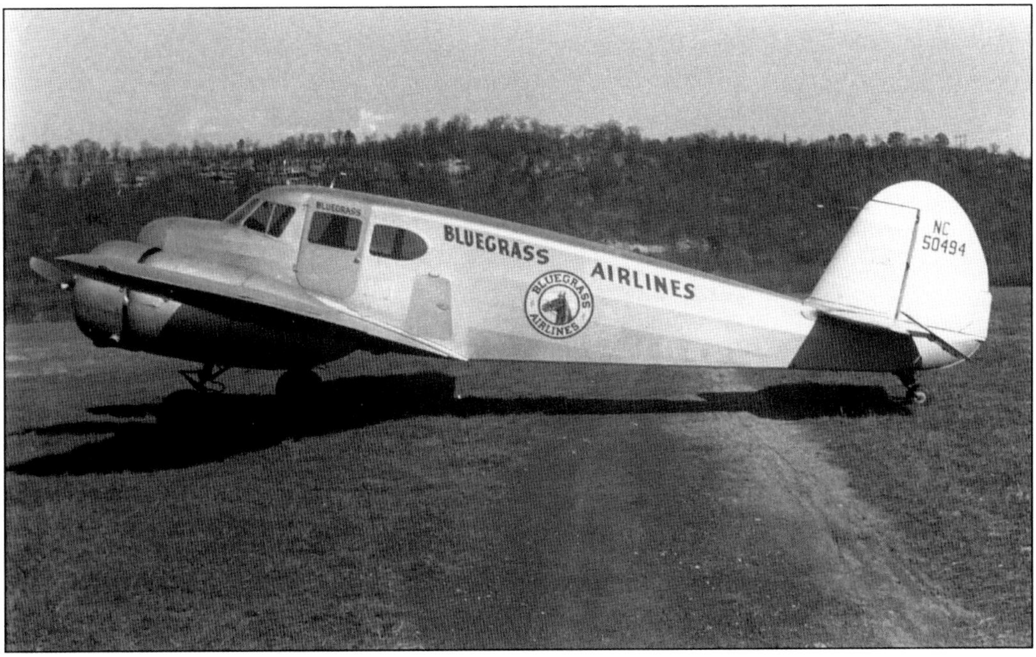

Starting operations in 1944, Bluegrass Airlines was a Kentucky intrastate operator that was headquartered in Russellville, Kentucky. In addition to Bowman Field, it provided passenger service to Paducah, Madisonville, Owensboro, Bowling Green, Lexington, and Ashland. The largest aircraft used were Cessna Bobcats (pictured above, but not at Bowman Field). Bluegrass Airlines was featured in the advertisement below from *Flying* magazine with a wonderful artist's rendering of a Bobcat at Bowman Field. Bluegrass Airlines ceased operations in 1947. (Above, Edward Peck; below, author's collection.)

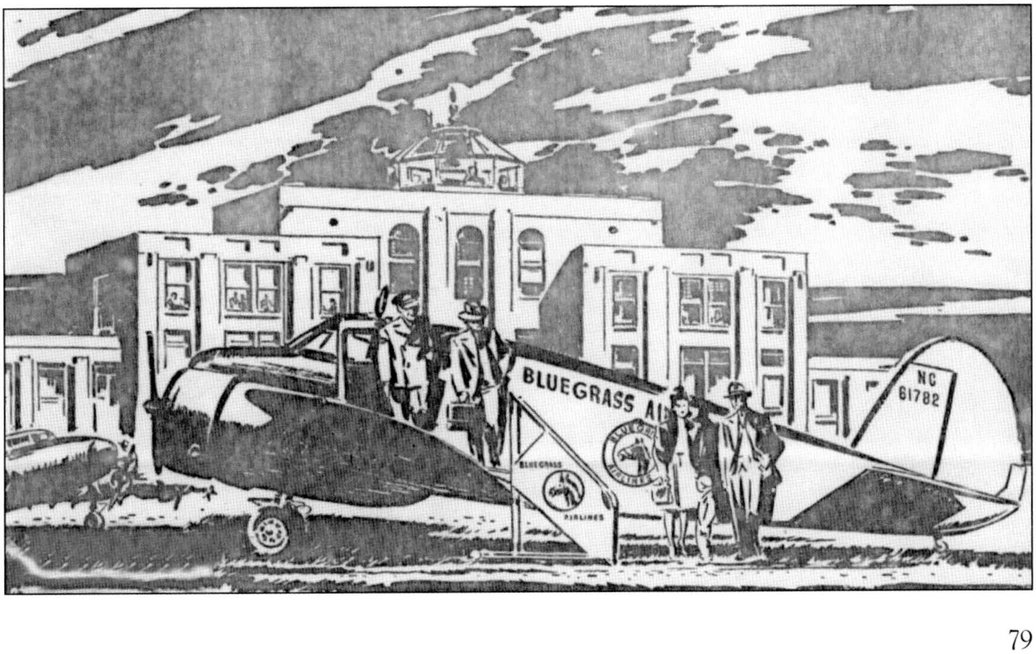

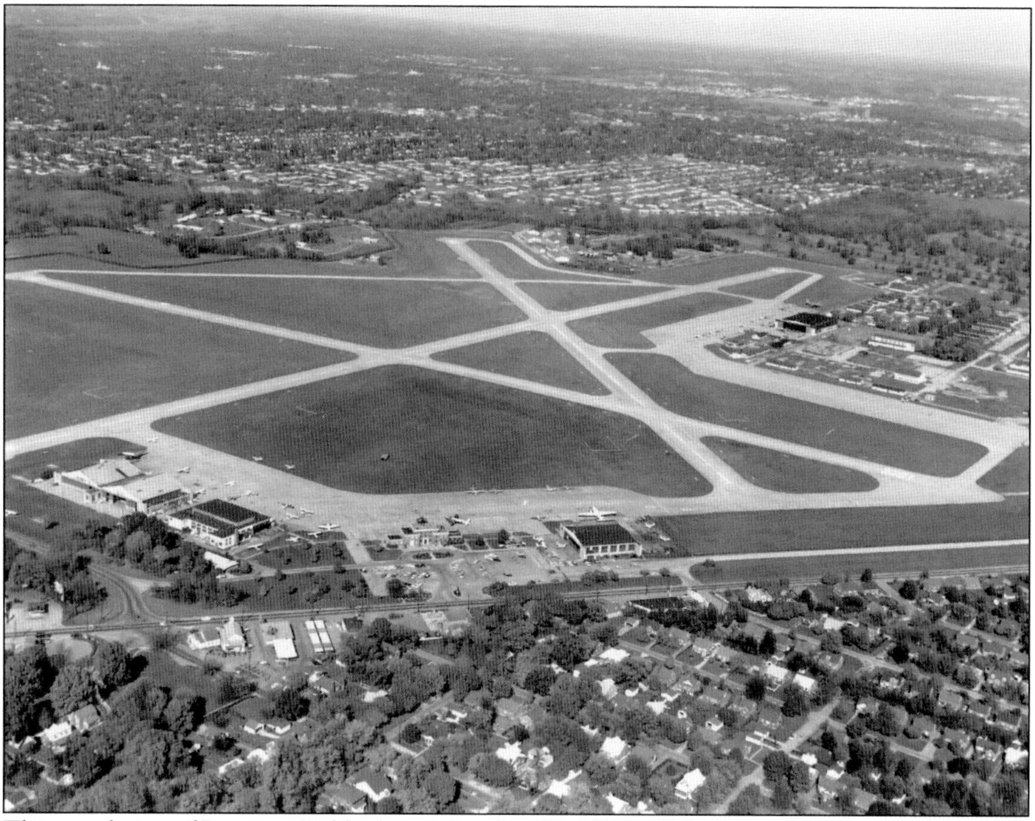

This aerial view of Bowman Field taken right after World War II shows how the surrounding areas had changed since the airport's beginnings. Residential developments are located to the northeast and south, while a country club is on the east along with a park and golf course to the north. Out of view to the west were more residential areas. All this nearby development led to neighborhood complaints of aircraft noise and low-flying airplanes, which continue to the present. (RAA.)

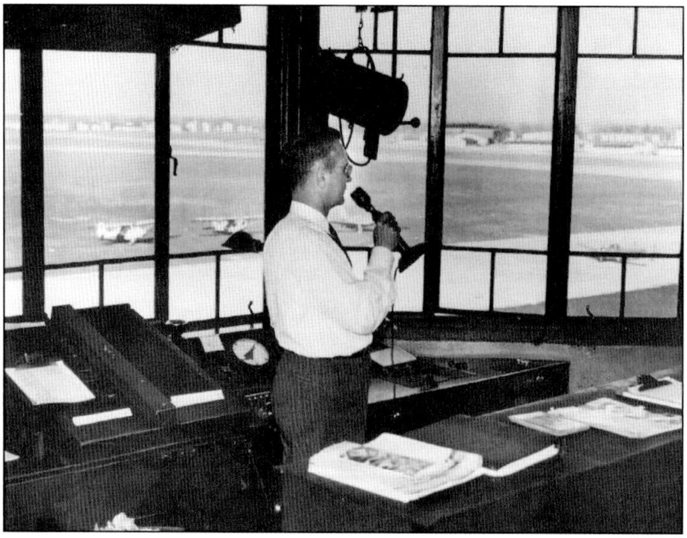

In the glassed-in structure atop the administration building, a Bowman Field air traffic controller uses two-way radio to talk to an airplane on the ground at the airport. A signal light above his head is used to communicate with airplanes that do not have a radio, and a wind direction indicator is located on the panel behind the controller. Fast forward 70 years, and air traffic control is still done about the same way, although nearly all airplanes now have a radio. (RAA.)

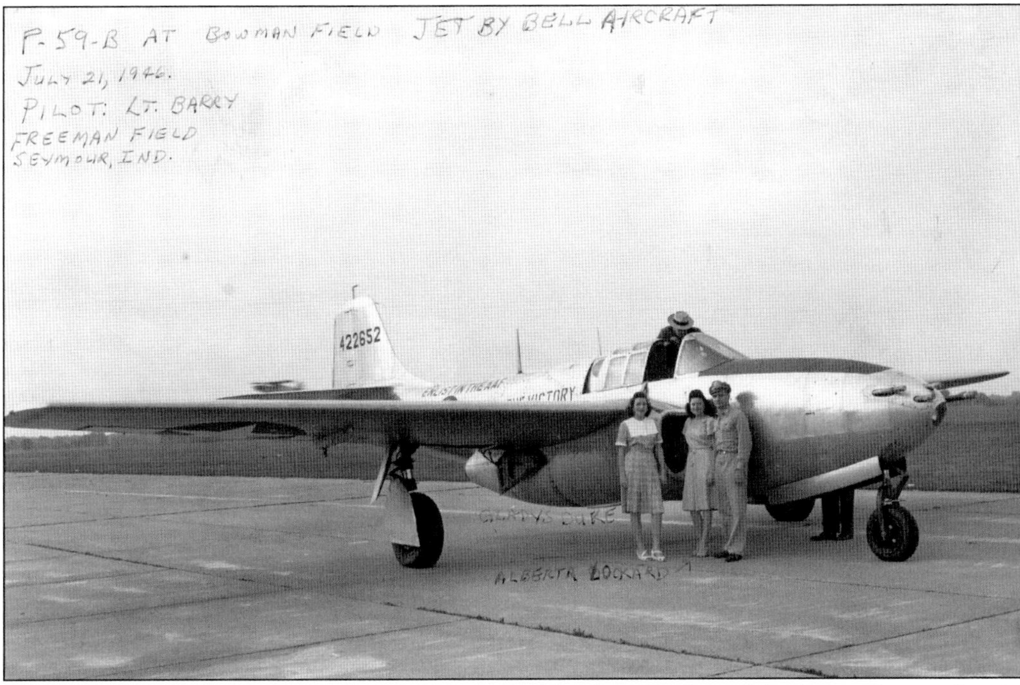

Although jet aircraft are common at Bowman Field today, the date of the first jet's arrival at the airport is uncertain. Nevertheless, this Bell P-59B jet at Bowman Field on July 21, 1946, could take the honors. Having flown in from Freeman Field in Seymour, Indiana, Lieutenant Barry poses with Gladys Duke (left) and Alberta Lockard. (Author's collection.)

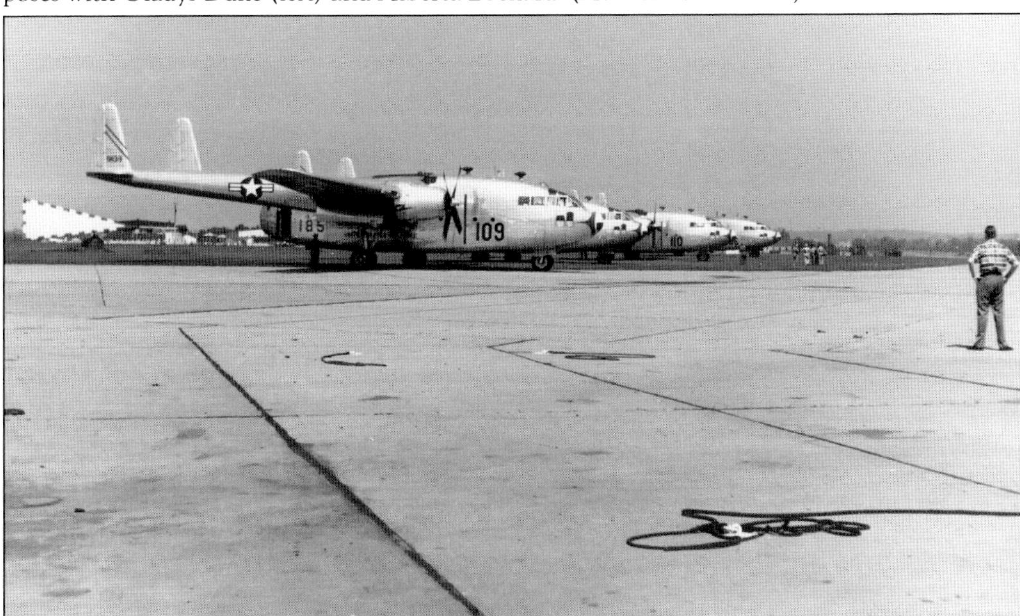

Several Fairchild C-119 Flying Boxcars visit Bowman Field during the early 1950s on what might be a cross-country training flight. The C-119 was a large troop and cargo carrying airplane. Nose markings indicate these aircraft are assigned to Seward Air Force Base in Symrna, Tennessee. Seward aircraft would serve in Korea during the Korean War. (Ed Taylor.)

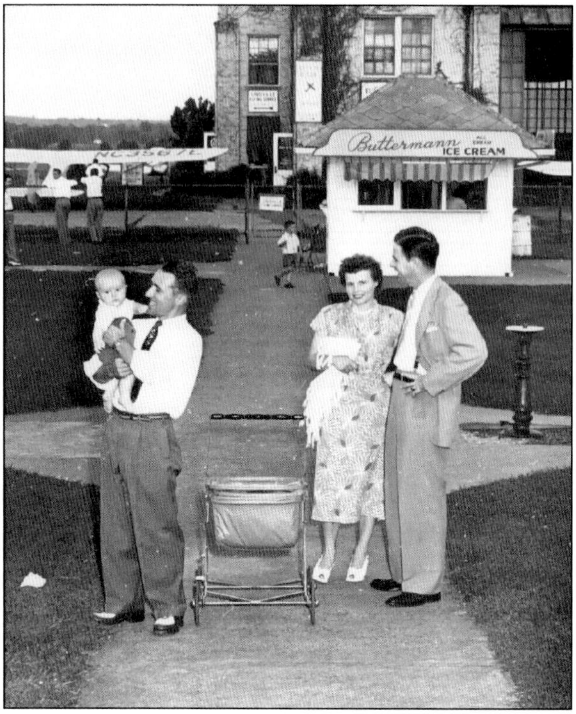

Forever known at Bowman Field as the "Popcorn Stand," this small structure dates to the 1930s and was still in use after World War II. In the 1949 photograph at left, the stand was operated by Boland's Restaurant. Jeannette and Bill Boland are standing at right. The same stand is pictured below in 1951, looking a little different, with Billy Boland Jr. showing Dolly Paris his popcorn plant growing in the crack between the concrete. (Left, author's collection; below, Wil Paris.)

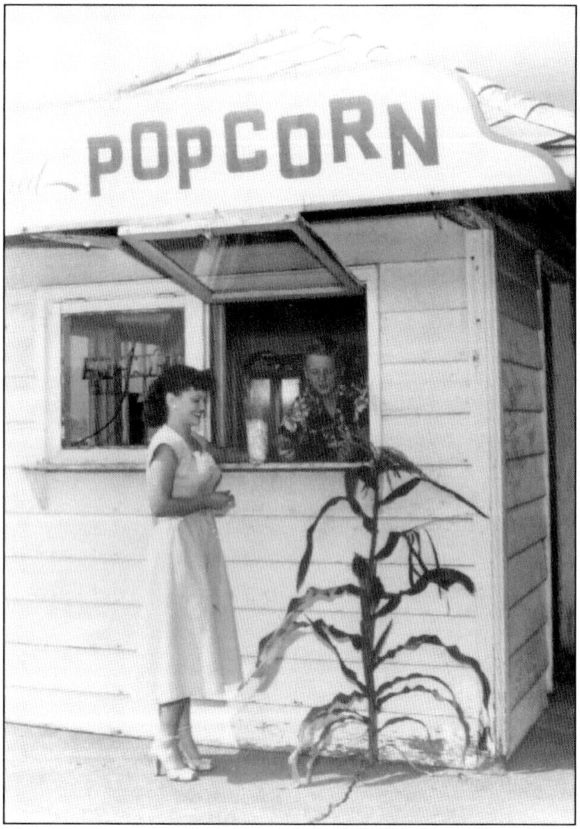

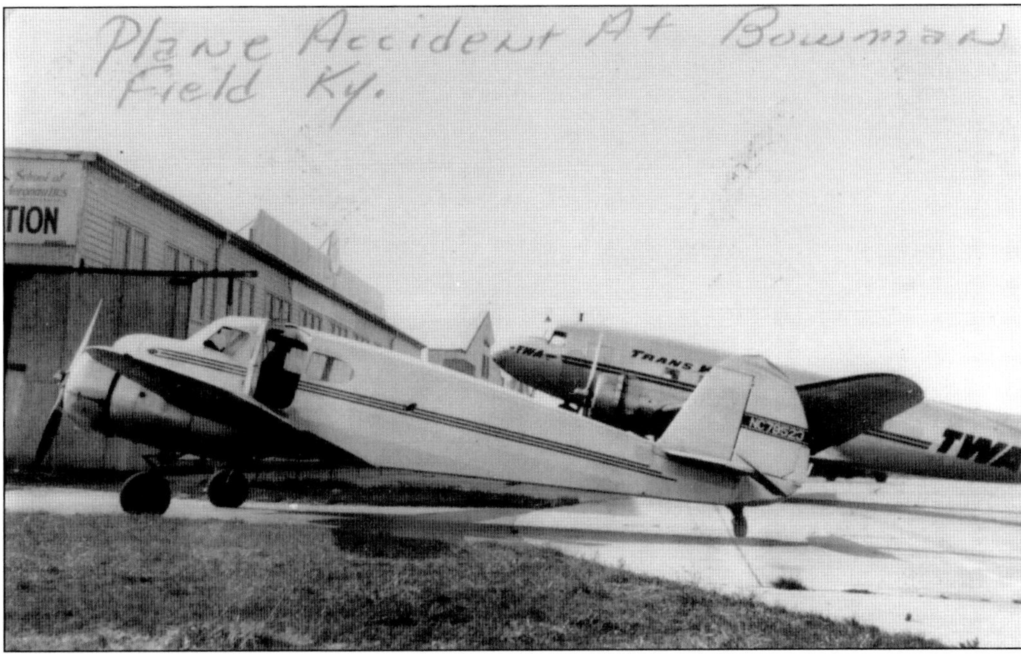

TWA was the last major airline to serve Bowman Field. Starting operations during June 1947 with eight daily flights, it was only at the airport until November, when all airlines moved to the new Standiford Field. Considering its short stay, photographs of TWA airplanes at Bowman Field are rare. This one shows a TWA DC-3 behind a Cessna Bobcat, which has a damaged tail from an accident. (Edward Peck.)

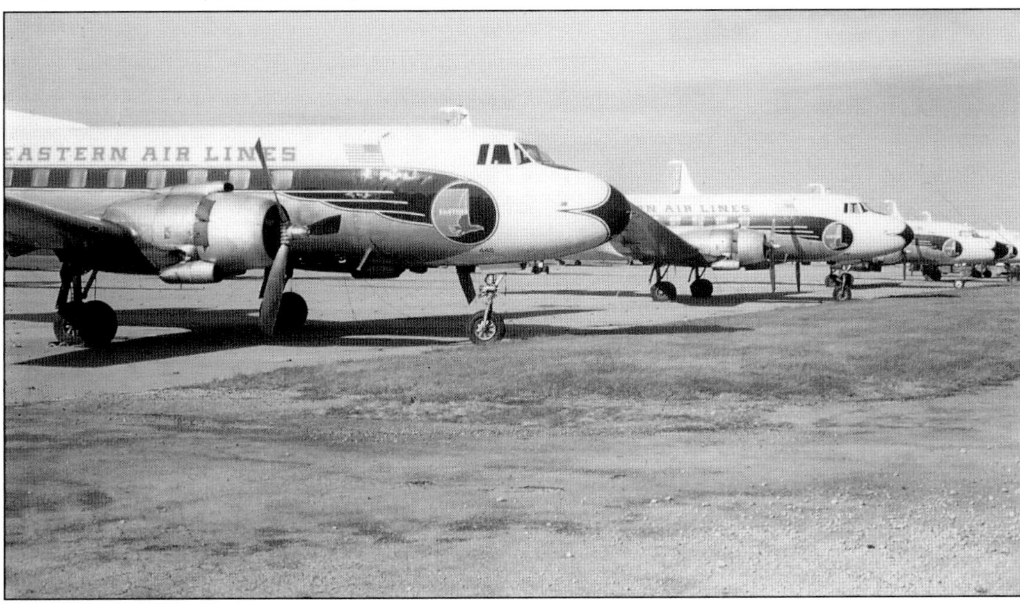

During the 1960s, Eastern Air Lines was plagued by numerous labor and management problems that resulted in strikes. At this time, fleets of Eastern Air Lines aircraft were parked at various locations while a strike was in progress. In 1962, Eastern has parked a number of its Martin 404 airliners on the east side of Bowman Field. Larger aircraft were sent to Standiford Field. (Charles Keeling.)

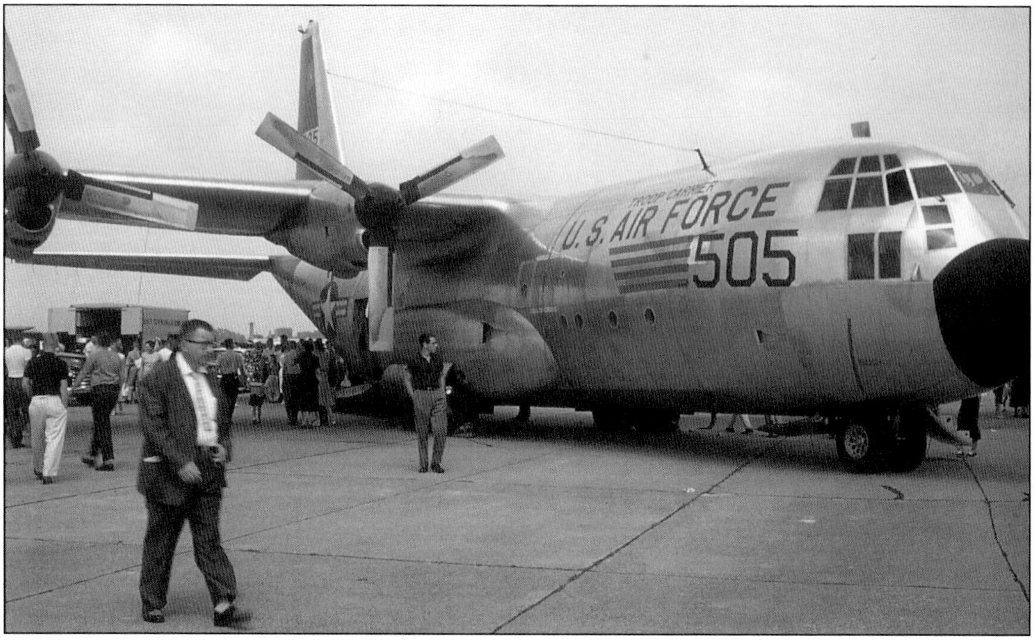

Sunday, June 15, 1958, was declared Bowman Field Appreciation Day. This was to celebrate what Bowman Field had meant to the city of Louisville, especially now that it was the city's primary general aviation airport and was setting records for takeoffs and landings. Appreciation day featured displays of aircraft for the public to view, including a Lockheed C-130 Hercules transport. The C-130 was likely the largest ever aircraft to visit Bowman Field. (Charles Keeling.)

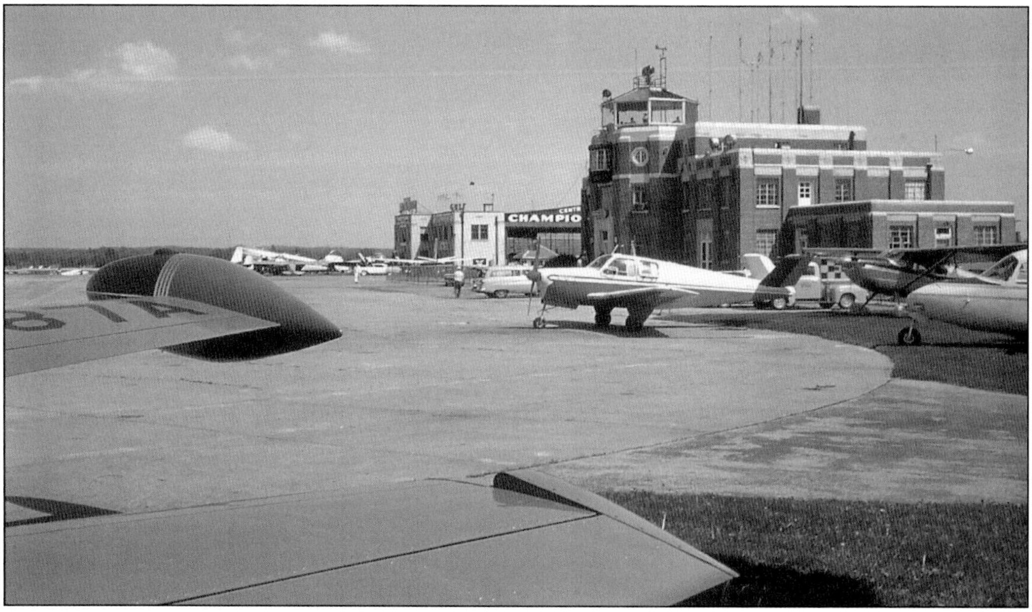

Bowman Field is seen in 1959, with airplanes scattered about the area of the administration building. Air traffic control operations are in the glassed-in structure above the building, which bristles with antenna to enable radio communications. This was a more relaxed time at the airport when a photographer could mingle with the aircraft and airport workers and not worry about the security restrictions of more recent times. (Charles Keeling.)

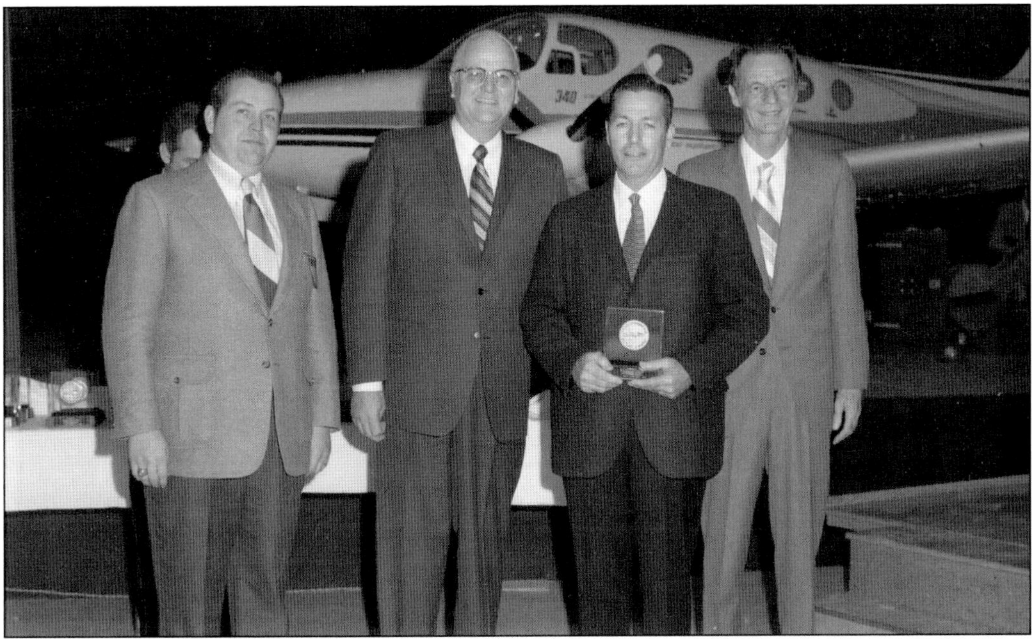

In the years following World War II at Bowman Field, two phrases were synonymous—Kentucky Flying Service and Dick Mulloy, pictured above second from right. Mentioning one would make someone automatically think of the other. During the Dick Mulloy years, Kentucky Flying Service was one of the busiest locations at Bowman Field and once had 120 airplanes in its environment. During this time, KFS was the place to be to learn to fly. Dick Mulloy and associates started the business during the late 1940s and moved to the big east side hangar in 1950. Believing in a personal relationship with his customers, Mulloy eventually bought out his partners. He retired in 1987 after 41 years. (Both, Mulloy family.)

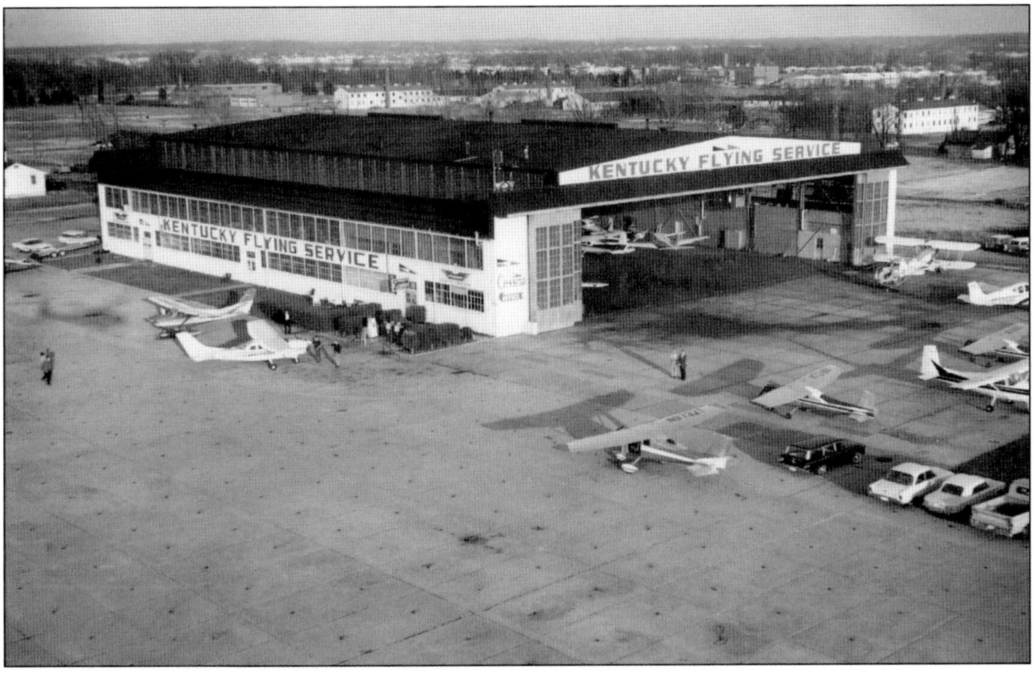

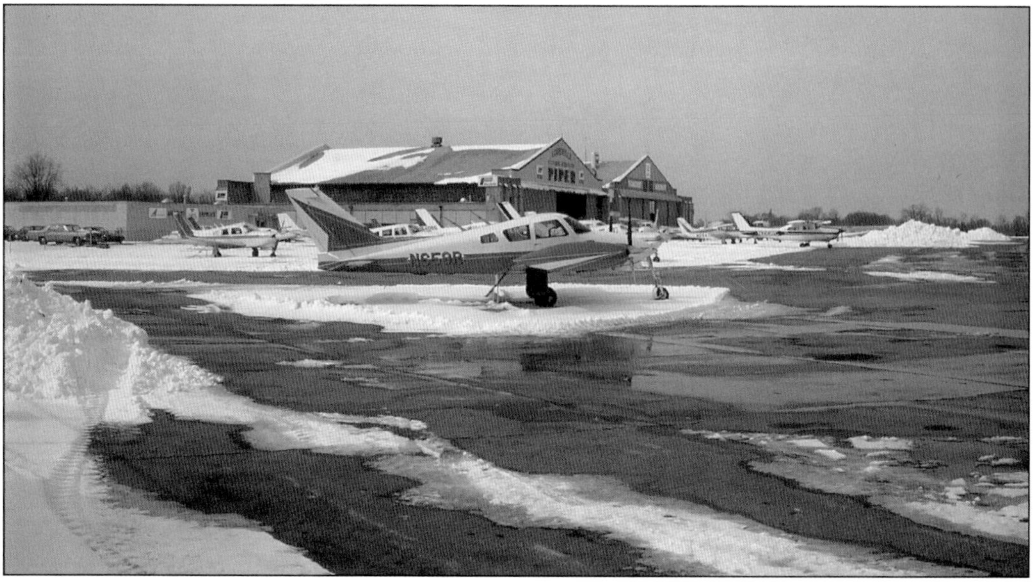

Louisville Flying Service, which had its start in 1932, continued in business after World War II. In 1963, its president and owner, John C. Bennett, was honored by friends and colleagues with the title "Mr. Aviation of Kentucky." Below, Bennett receives a trophy with his new title. The others are, from left to right, Charles R. D'Olive, William Piper of Piper Aircraft, Mike Murphy, and Leonard Lusky. In 1962, Bennett marked 25 years with Louisville Flying Service. (Above, author's collection; below, Edward Peck.)

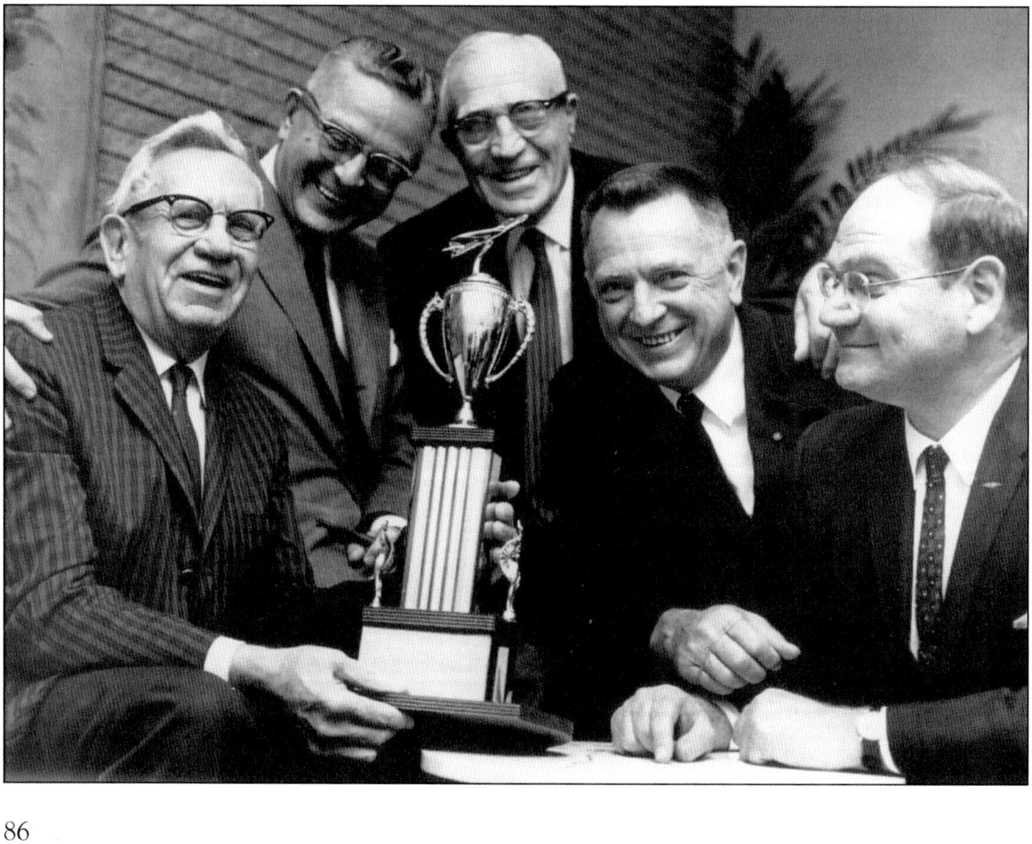

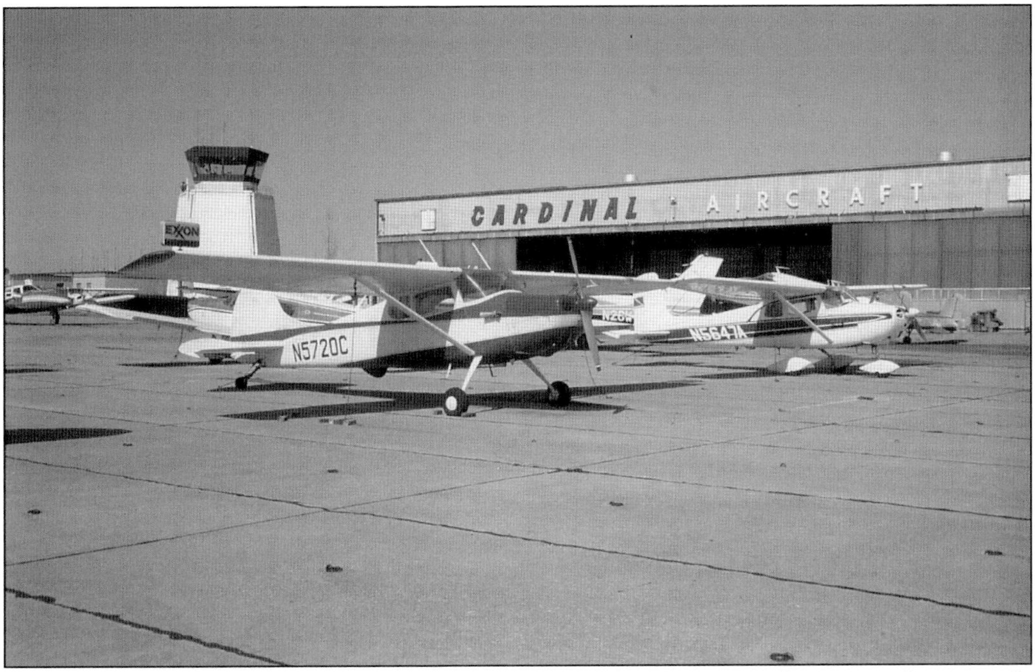

Flying service Cardinal Aviation opened a brand new hangar in April 1966 on the east side of Bowman Field, just west of Kentucky Flying Service. Actually, it was the former Kentucky Air Transport (KAT) that took a new name with the new building. KAT was originally in the wartime engineering building that was located between the administration building and the twin hangars. Cardinal had J.B. Wathen as president and Al Seelbach as manager. KAT was originally incorporated in 1944. (Author's collection.)

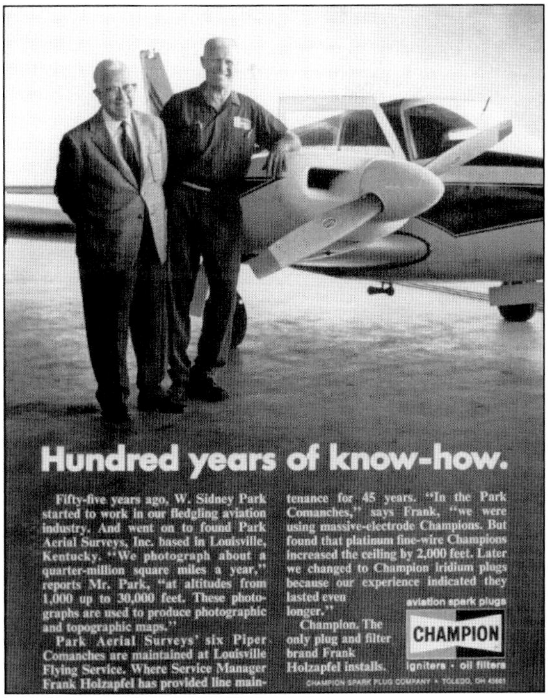

Champion, which was a major manufacturer of aircraft spark plugs, singled out two Louisville aviation notables in this magazine advertisement. Wallace Sidney Park, Abram Bowman's second partner, was a national leader in aerial photography and the owner of Park Aerial Survey. Frank Holzaphel was co-owner of Louisville Flying Service and a renowned aircraft mechanic. This ad was quite appropriate considering that Louisville Flying Service hangared and serviced Park's airplanes. (Author's collection.)

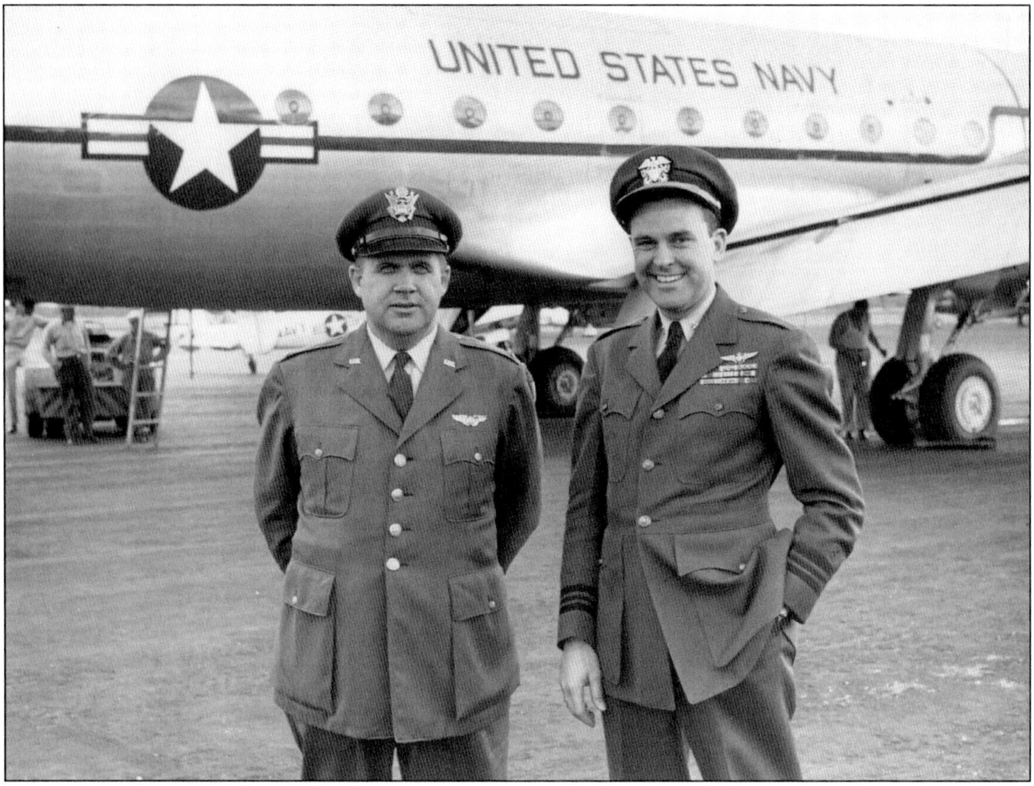

Central American Airways traces its roots at Bowman Field to 1946, when it started operations from one of the World War II buildings on the east side of the field, moving to its present location in 1950. Founded by W.L. "Wil" Paris, Wally Miller, and Bill Stuber, in time, it became an all-Paris business with wife Dolly and now son W.L. as proprietors. In earlier years, CAA provided non-scheduled airline service from Bowman Field, maintained large aircraft for cargo and passenger operations, and held a US Army contract to train military pilots. During the 1950s, Paris (above, right) commanded Naval Reserve Transport Squadron VR-692 based in Columbus, Ohio. (Above, author's collection; below, Wil Paris.)

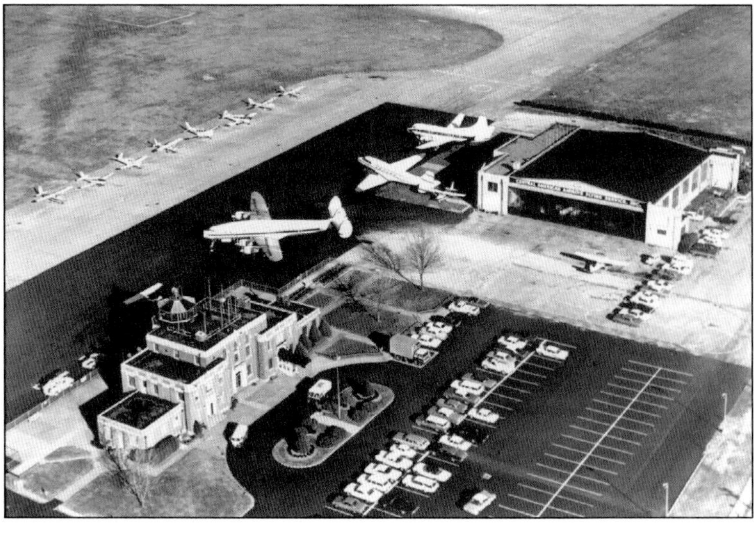

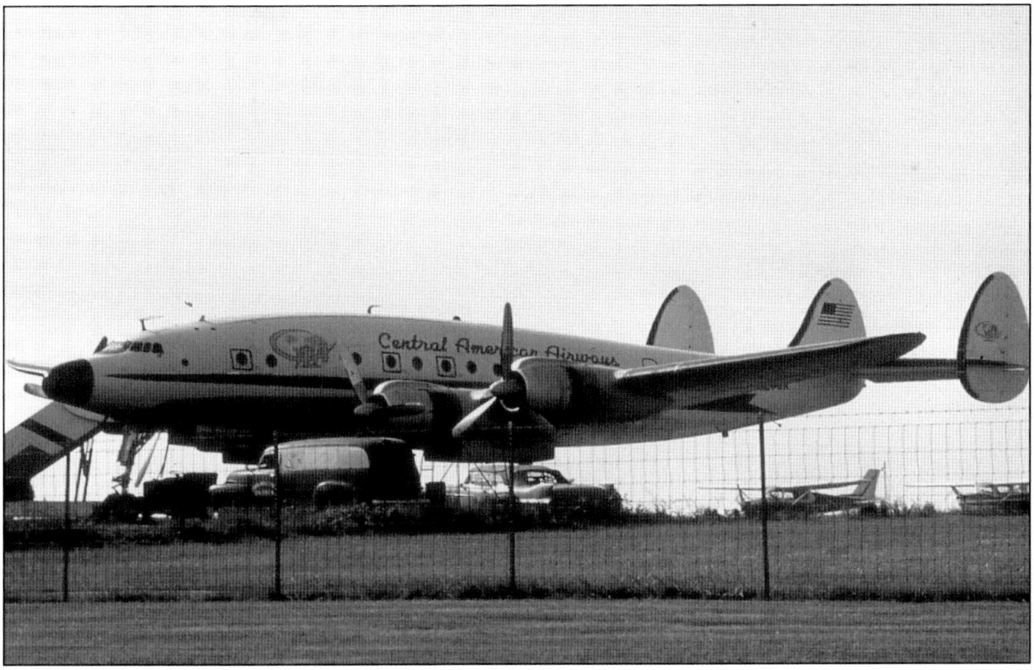

Into the 1980s, Central American Airways maintained several large aircraft at Bowman Field to carry passengers and freight. They included a Lockheed 749 Constellation and a Martin 202. The Martin was used to fly local sports teams and was often chartered by touring rock bands. Central American also had several Douglas DC-3s at one time or another, and in the late 1940s, it offered nonscheduled airline service to New York City from Louisville. Current charter service uses Cessna Citation jets and Beechcraft twin-engine prop aircraft. (Both, Charles Thornton.)

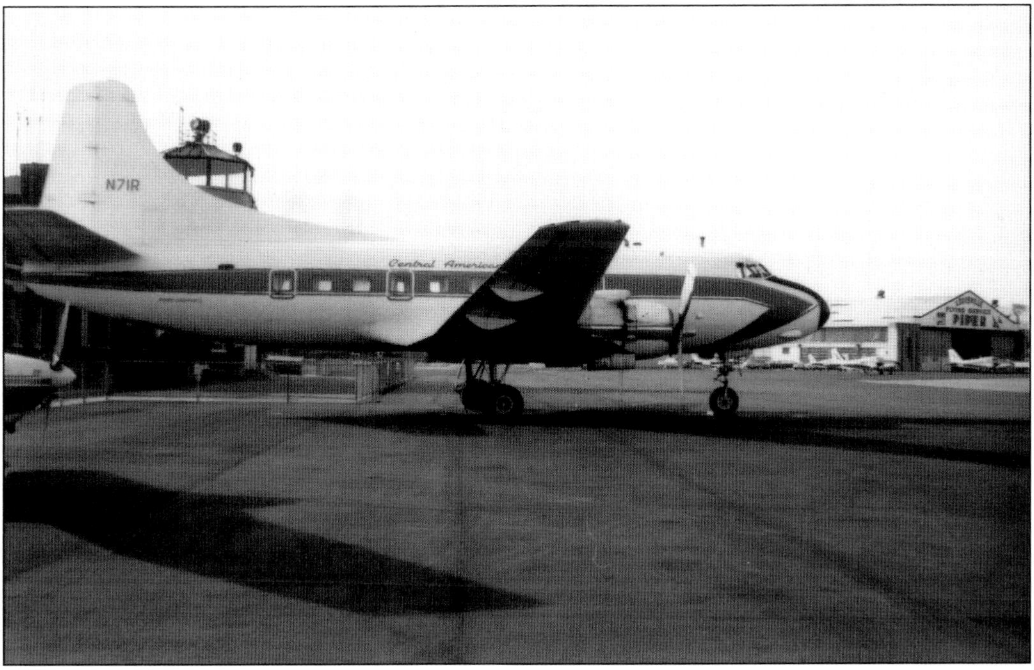

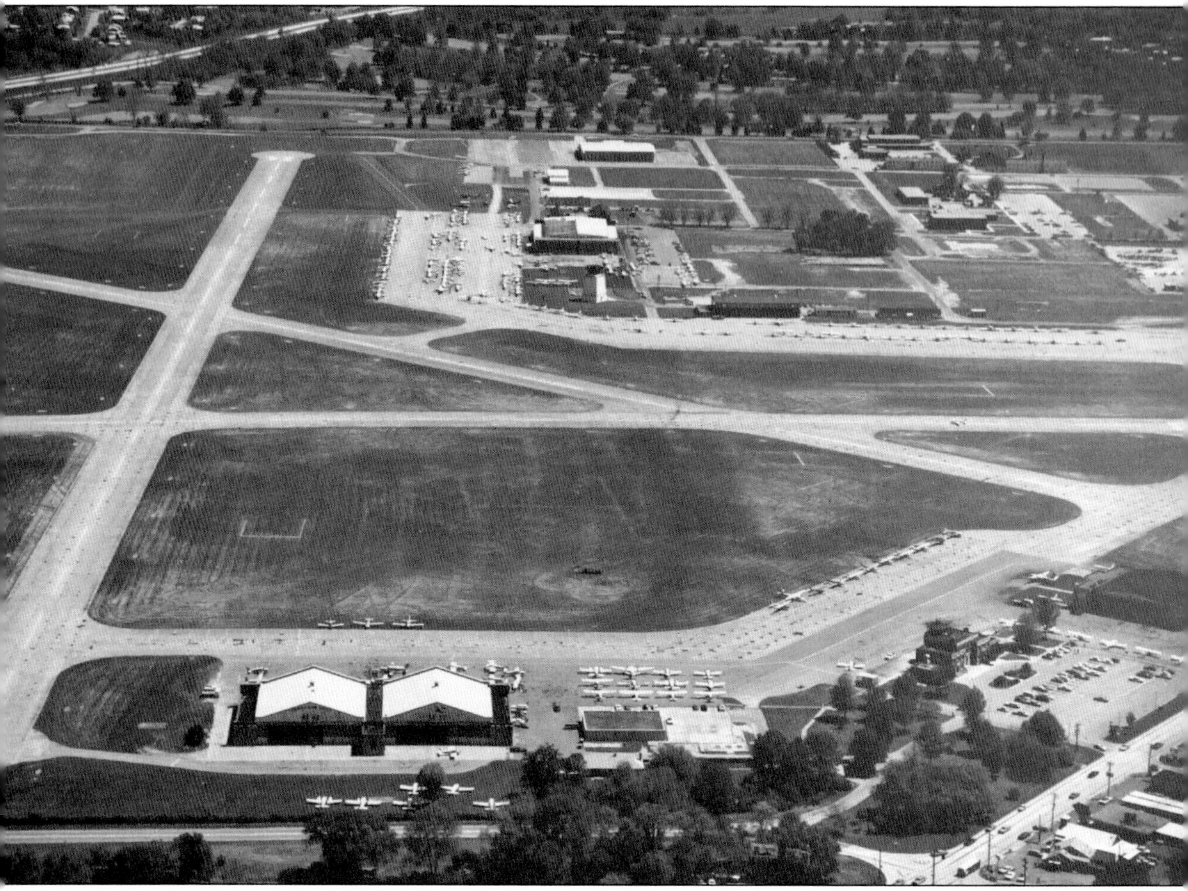

In 1958, Bowman Field was acknowledged as the busiest local traffic airport in the United States. This 1970 photograph clearly shows the reason for this. Busy airports have lots of airplanes, as seen here, which can generate lots of flights. Another clue is the half square seen in the grassy area to the right of the east-west runway. This would enable three airplanes to land on the same runway at the same time, with one on the hard surface and one on each side in the grass positioning with the half square. (Ed Taylor.)

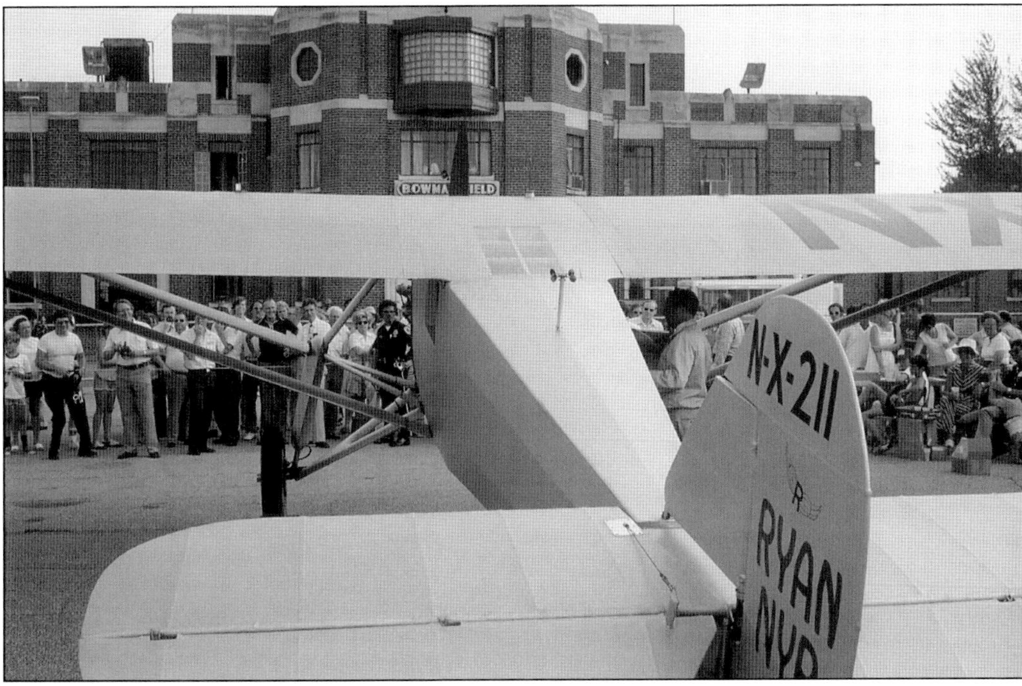

To celebrate the 50th anniversary of Charles Lindbergh's transatlantic flight to Paris, the Experimental Aircraft Association put its *Spirit of St. Louis* replica on a national tour in 1977. It visited Bowman Field on July 6 and was admired by a large crowd of people. Below, dressed like Lindbergh, pilot Vern Jobst poses with a group of Civil Air Patrol cadets underneath the wing of the *Spirit*. (Both, author's collection.)

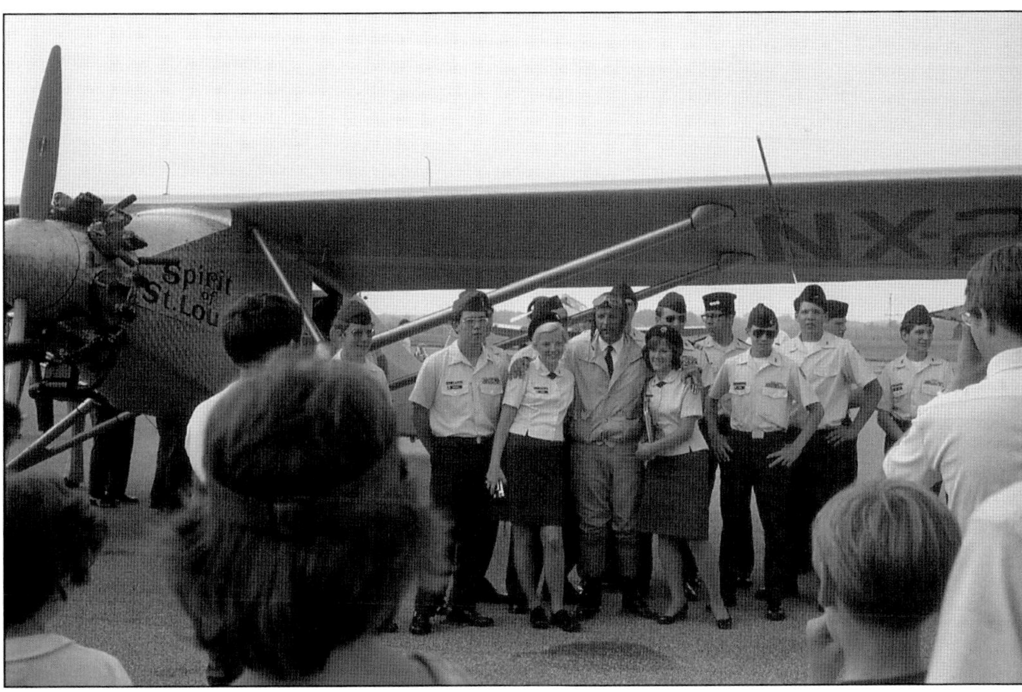

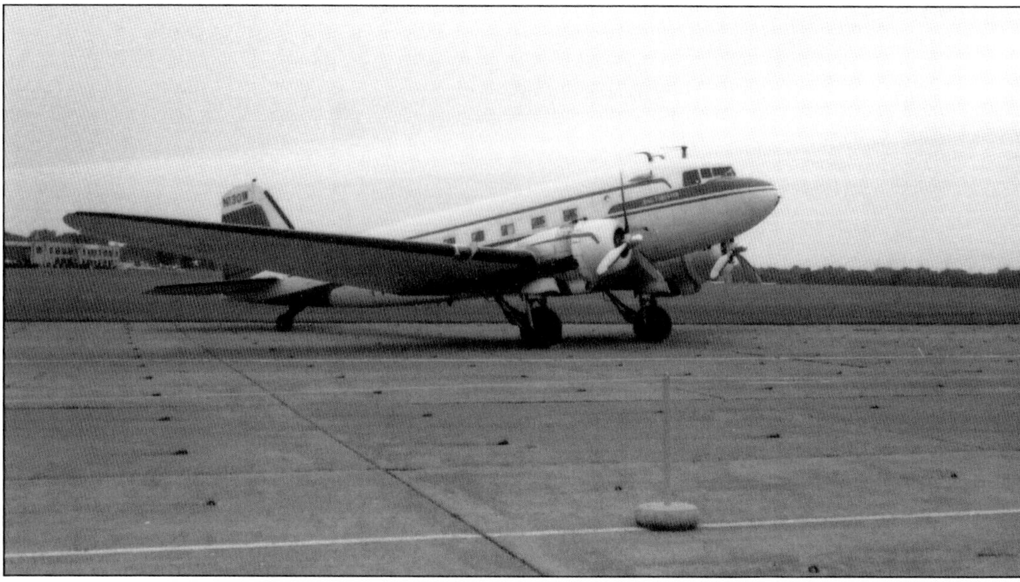

Over the years, numerous corporate aircraft have been hangared at Bowman Field. One of the post–World War II leaders was Brown-Foreman Distillers, which entered corporate aviation in 1954 with a Beech 18 and quickly replaced it with two Douglas DC-3s. In 1964, Brown-Foreman brought the first corporate jet airplane to Bowman Field with the acquisition of a North American Sabreliner. Registered N30W, the "W" stood for "Whiskey," the same letter used to end all Brown-Foreman aircraft registrations. (Both, author's collection.)

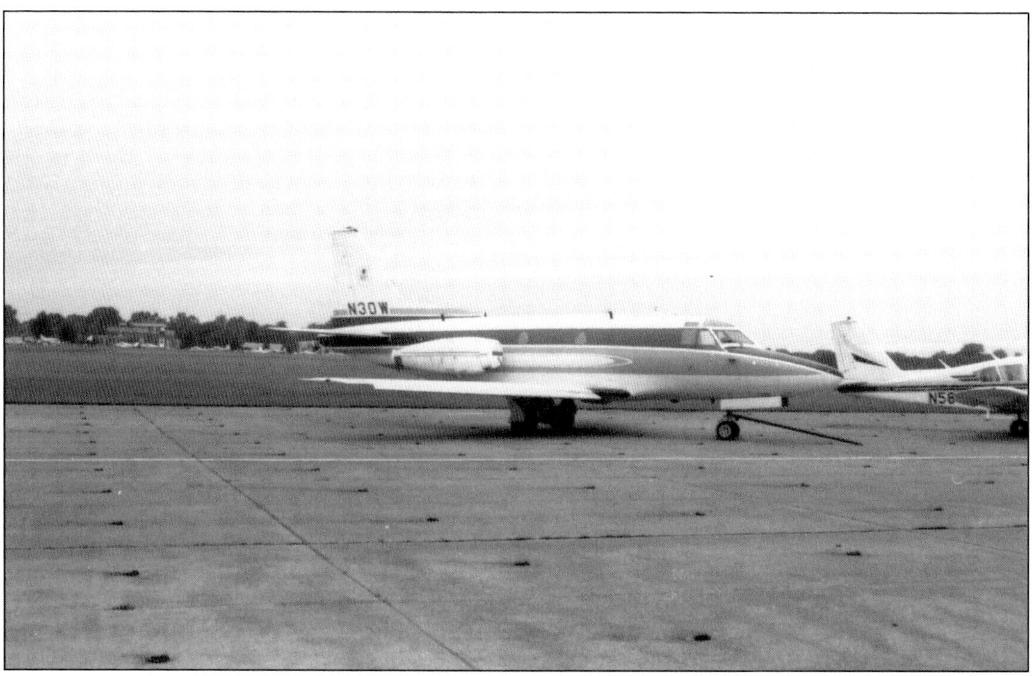

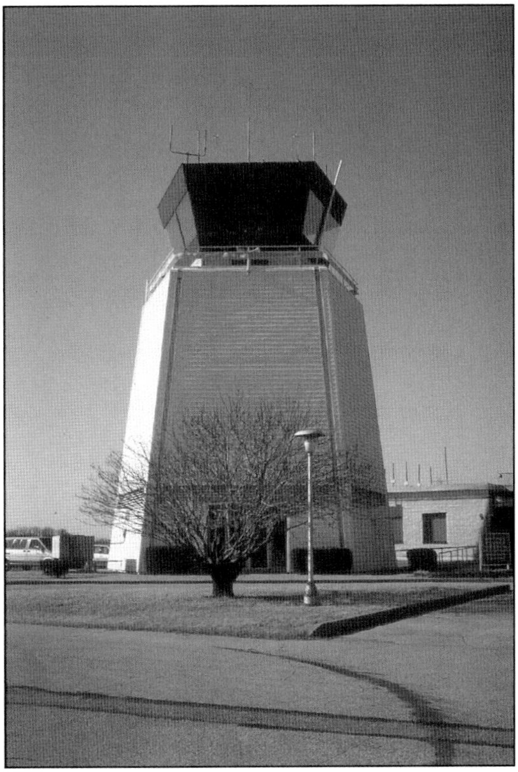

Air traffic control has always been important at airports and became significant after World War II with the popularity of small general aviation aircraft. At Bowman Field, air traffic control was first conducted from the glassed-in area above the administration building and was transferred to a newly built control tower dedicated on July 17, 1965. This tower, which uses two-way radio to communicate with aircraft, was placed on the east side of the field so operators could have a better view of the airport. (Both, author's collection.)

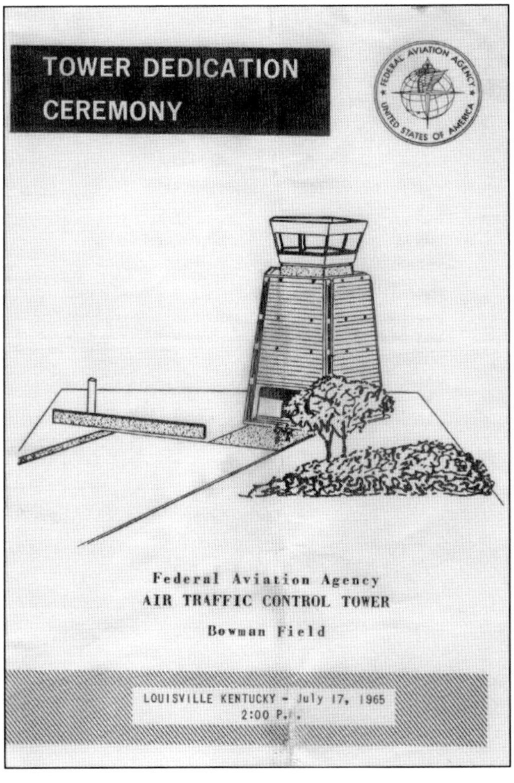

TOWER DEDICATION CEREMONY

Federal Aviation Agency
AIR TRAFFIC CONTROL TOWER
Bowman Field

LOUISVILLE KENTUCKY - July 17, 1965
2:00 P.M.

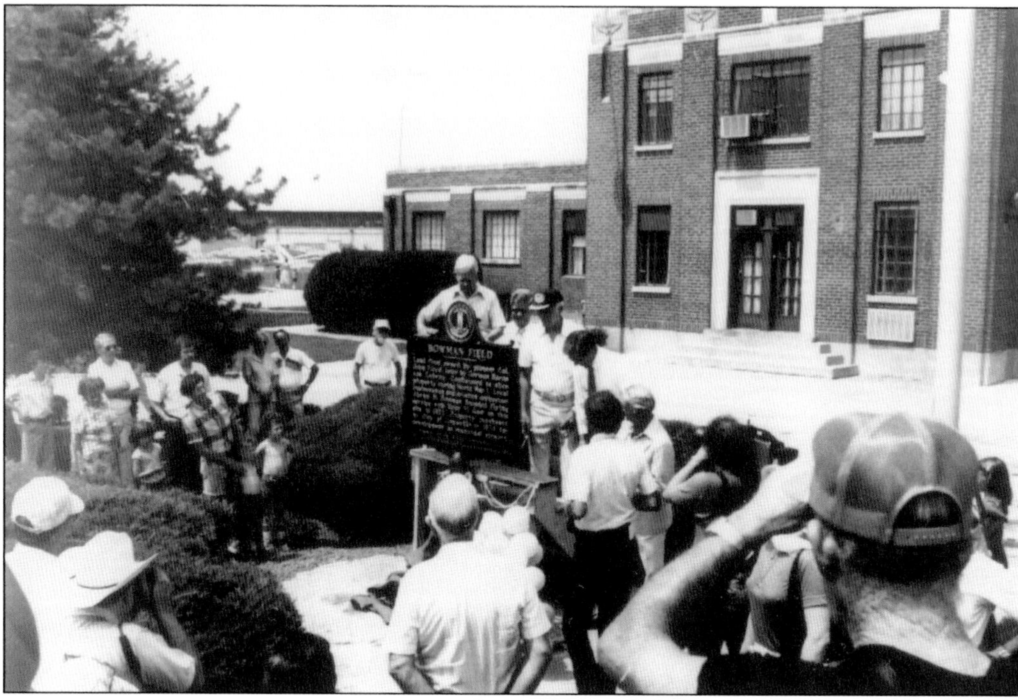

During a 1980 ceremony, a Kentucky state highway marker was dedicated at Bowman Field to commemorate the history of the airport. A short history inscribed on the marker noted the original ownership of the land and the airport's founders. Currently, two other highway markers are on airport property. Several historical aircraft including a Stinson A Tri-Motor were on hand to help celebrate the dedication. (Above, Edward Peck, below, author's collection.)

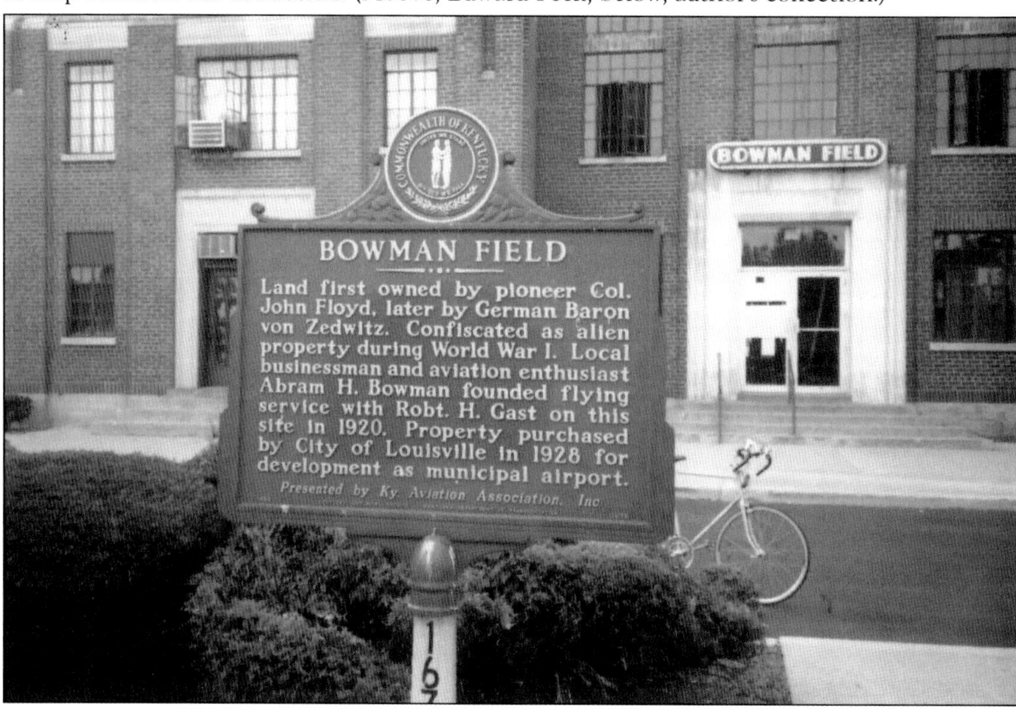

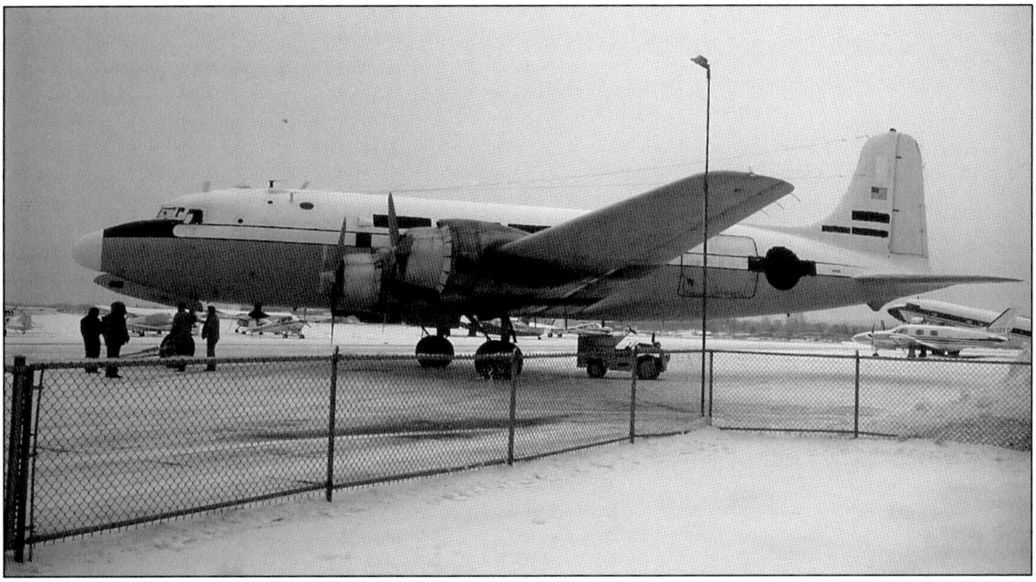

During the years after World War II, sighting of large aircraft at Bowman Field was quite unusual. This changed on a wintry January 11, 1979, when a former military Douglas C-54 four engine aircraft landed during the afternoon and was abandoned. Investigation revealed that the airplane was carrying a load of illegal marijuana that was unloaded in Lexington. The plane was then flown to Bowman Field, where the crew disappeared in a waiting vehicle. (Author's collection.)

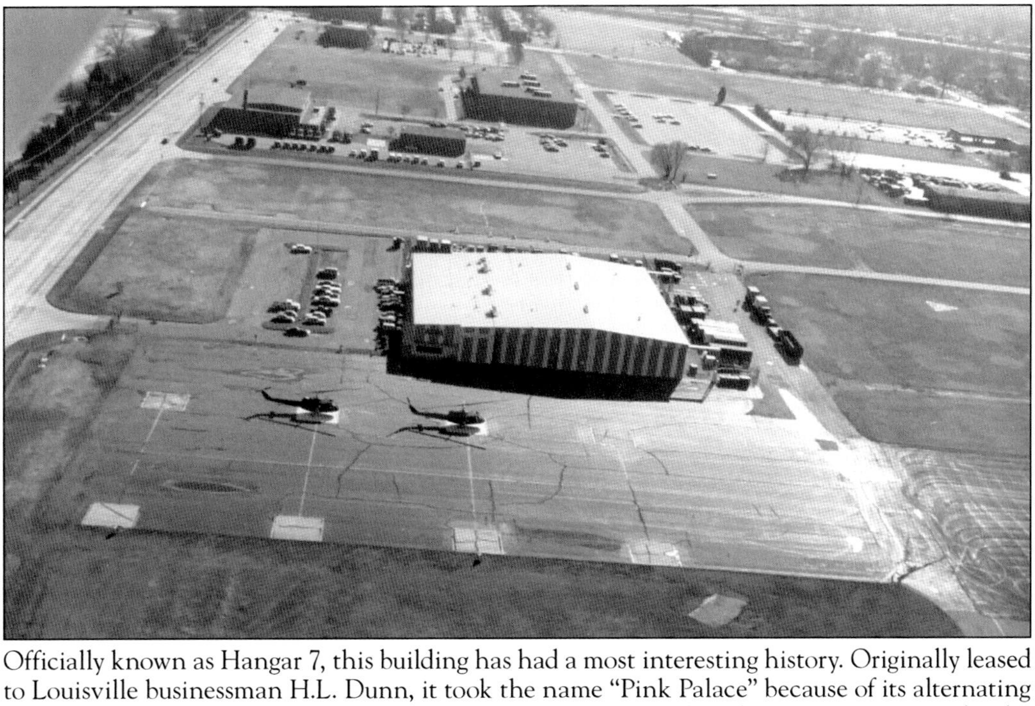

Officially known as Hangar 7, this building has had a most interesting history. Originally leased to Louisville businessman H.L. Dunn, it took the name "Pink Palace" because of its alternating pink and white stripes. It is best remembered at Bowman Field as the operation center for the US Army Reserve 412th Medivac Helicopter Squadron. The 412th was at Bowman Field from 1975 to 1989. Recently, the hangar has had an educational role serving as the home of the Air + Space Academy. (Author's collection.)

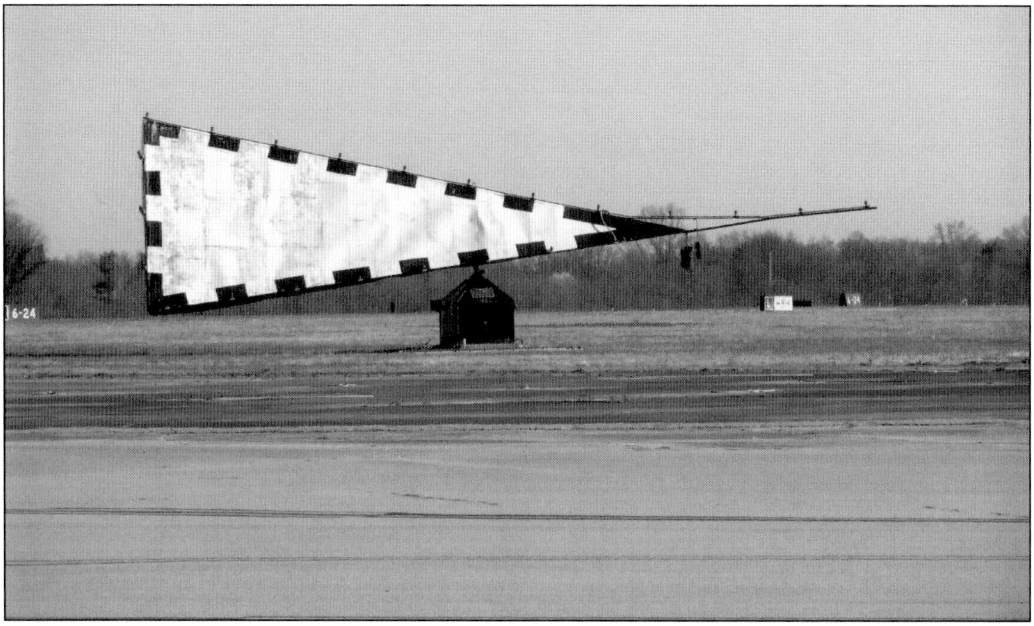

A longtime fixture at Bowman Field is the tetrahedron, which is located near the administration building. Installed during the late 1930s, it is used to indicate wind direction. It is still functioning and was recently overhauled by members of the Kentucky Aviation History Society. The overhaul included repair, repainting, and replacement of incandescent light bulbs with energy-saving LEDs. (Author's collection.)

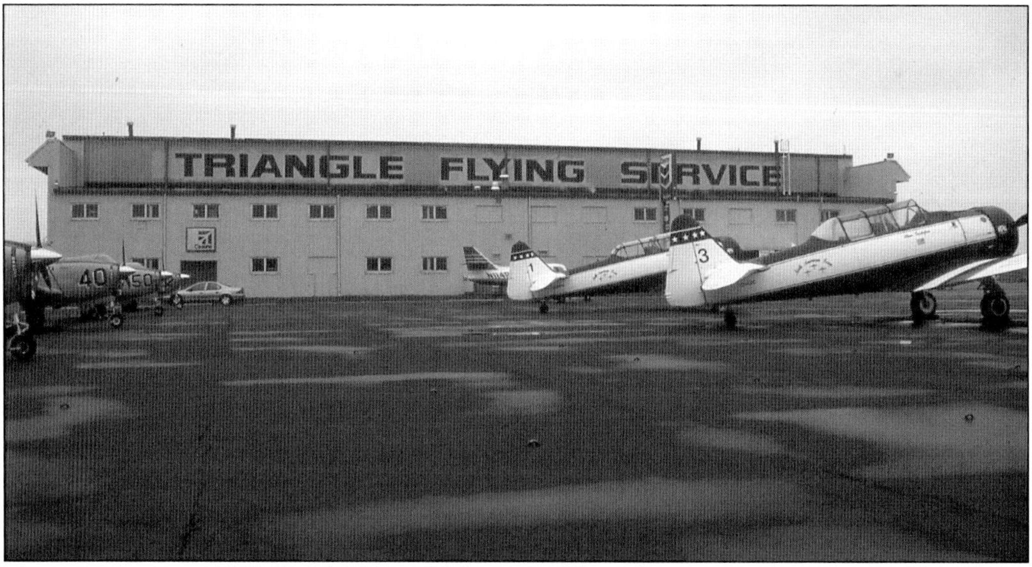

Upon retirement in 1987, Dick Mulloy sold Kentucky Flying Service to new owners Jack Korfhage, John Clark, and Lou Meiners. They maintained the service until June 1994, when they closed the business due to a rent dispute with the Regional Airport Authority. To keep a flying service on the east side of the airport, Central American Airways was appointed interim operator. In September, Triangle Flying Service was selected to be the proprietor. Triangle, which was based in Campbellsville, Kentucky, began operations on October 1. The service provided fuel and aircraft repair. (Author's collection.)

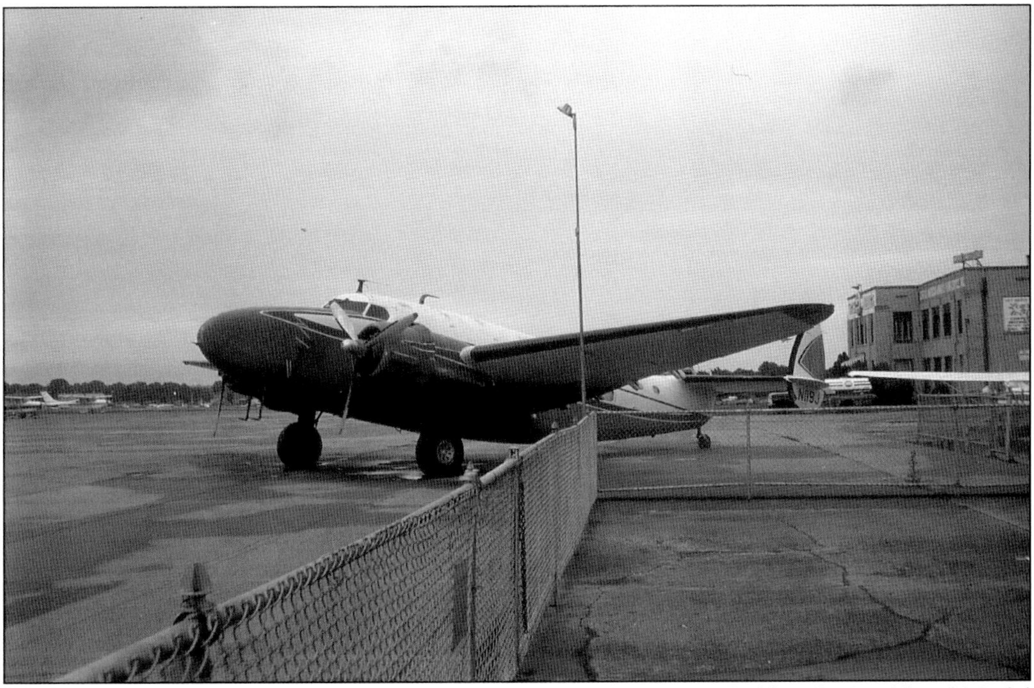

Perhaps the most talked-about airplane in the history of Bowman Field is the Lockheed Model 18 Lodestar, which is pictured here parked on the west side of the airport. Rumors abound that it was an abandoned drug carrier, a lost airliner, Howard Hughes's airplane, and many more. Actually, the airplane was purchased by Central American Airways with the intention to reuse its engines by installing them on other CAA aircraft. The big Lockheed arrived in 1973 and was still at Bowman Field in 2017. (Both, author's collection.)

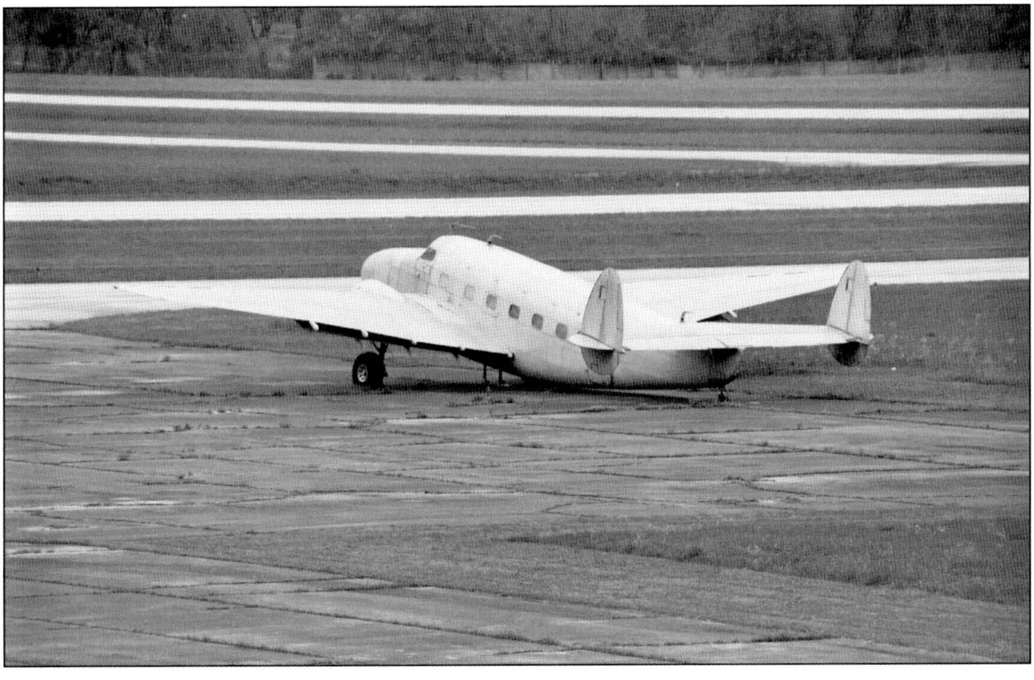

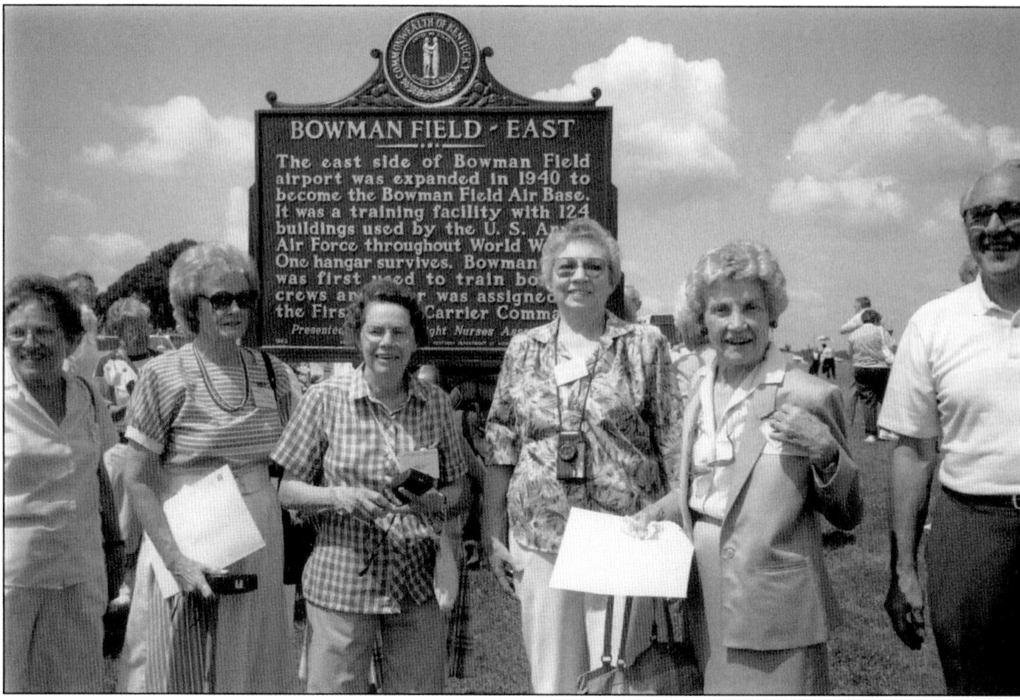

In June 1992, the World War II Flight Nurses Association held its convention in Louisville. Over 100 former flight nurses attended, and they made a return to Bowman Field, where most of them were trained. At this time, a Kentucky state highway marker was dedicated that paid tribute to the School of Air Evacuation. Attending was Ethel Guffey Simpson (second from right) along with her husband (right). (Both, author's collection.)

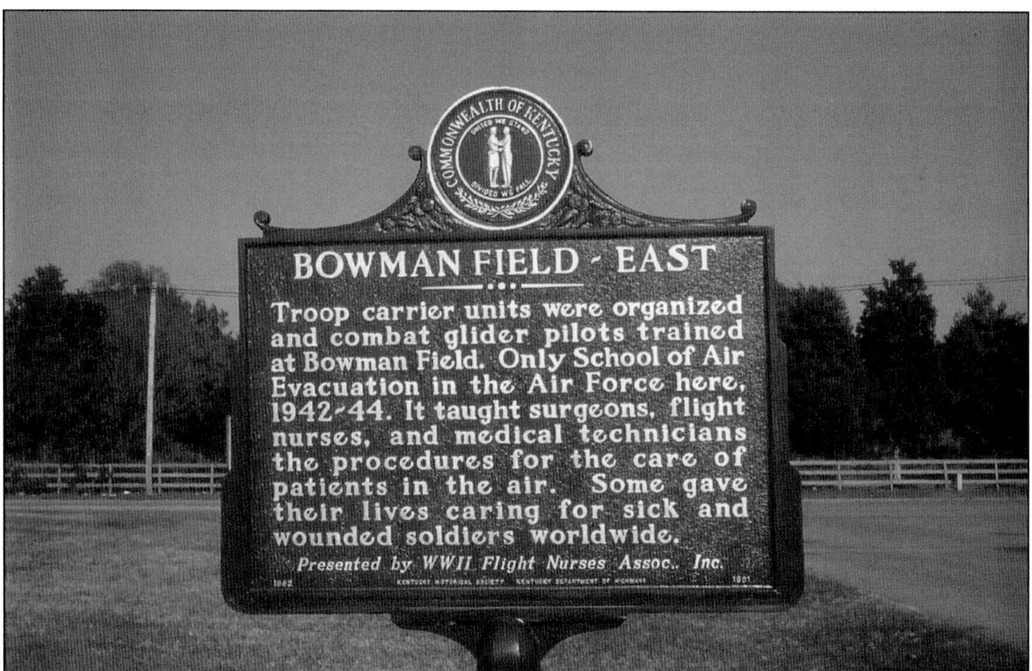

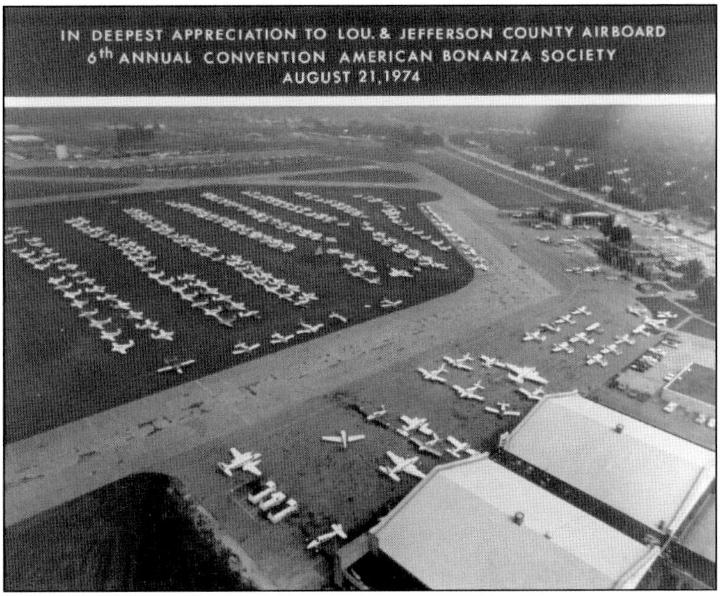

Special events have always been important at Bowman Field. One of the biggest was the gathering of the Beech Bonanza Society during 1974. The many participants required much of the airport's grassy center area for parking. The Bonanza has long been recognized for its quality and performance while early models featured a "V" tail. New ones, although only available with a standard tail, are still being produced. (Author's collection.)

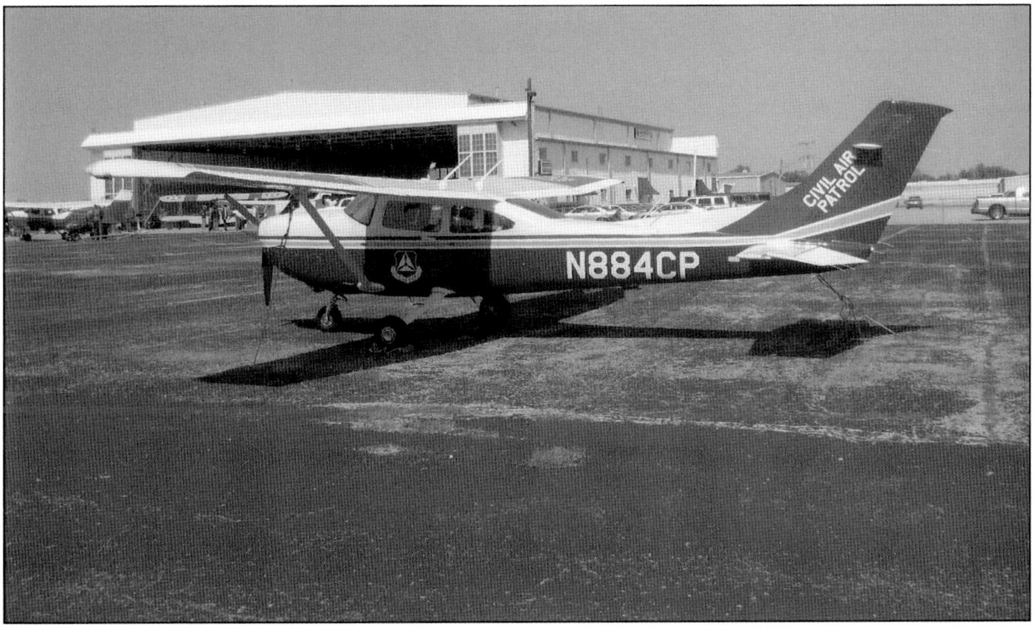

The Civil Air Patrol (CAP) was founded in the months before World War II as an auxiliary to the Air Force. Squadrons were assigned to each state including Kentucky. From the beginning, Bowman Field has had a CAP presence. Currently the airport has a composite squadron that includes both senior and cadet members. The CAP is tasked with searching for lost, missing, or downed airplanes and assisting during periods of emergency. Youths are encouraged to join and participate in assigned missions. Presently, the CAP at Bowman Field flies Cessna high-winged aircraft. (Author's collection.)

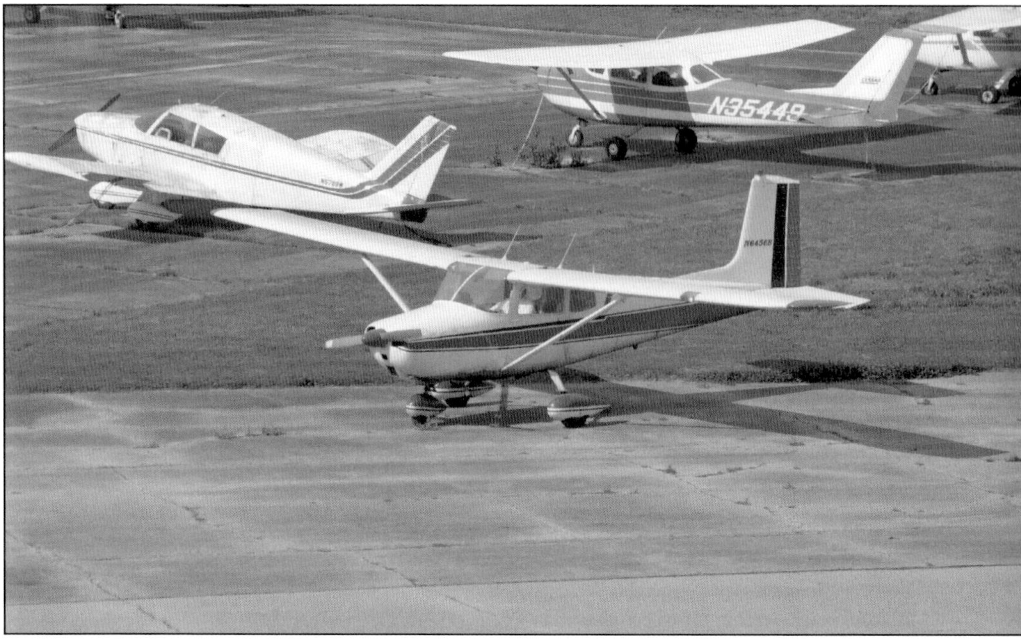

Over the years, the most common aircraft seen at Bowman Field would have to be the Cessna 172. Introduced during the mid-1950s, over 34,000 have been made, and it is still in production. Many older Cessna 172s are still flying, including this 1956 model that visited Bowman Field in 2017. (Author's collection.)

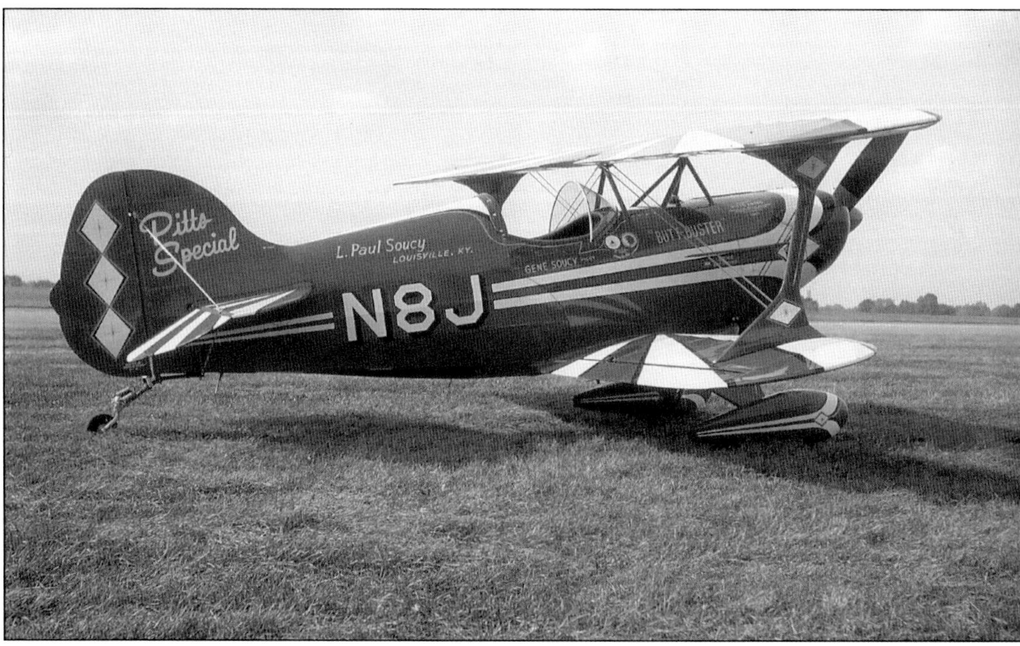

An alumni list of Bowman Field trained pilots is endless. One of the best known is Gene Soucy, who travels the national air show circuit performing his aerobatic routine. Soucy began flying at Kentucky Flying Service during the 1960s and learned aerobatics in his father, Paul's, Pitts Special, N8J. Named "Mr. Air Show," Gene Soucy has appeared in over 3,000 air shows and was a member of two world champion US aerobatic teams. (Charles Keeling.)

Five

Air Shows, Fly-Ins, and Festivals

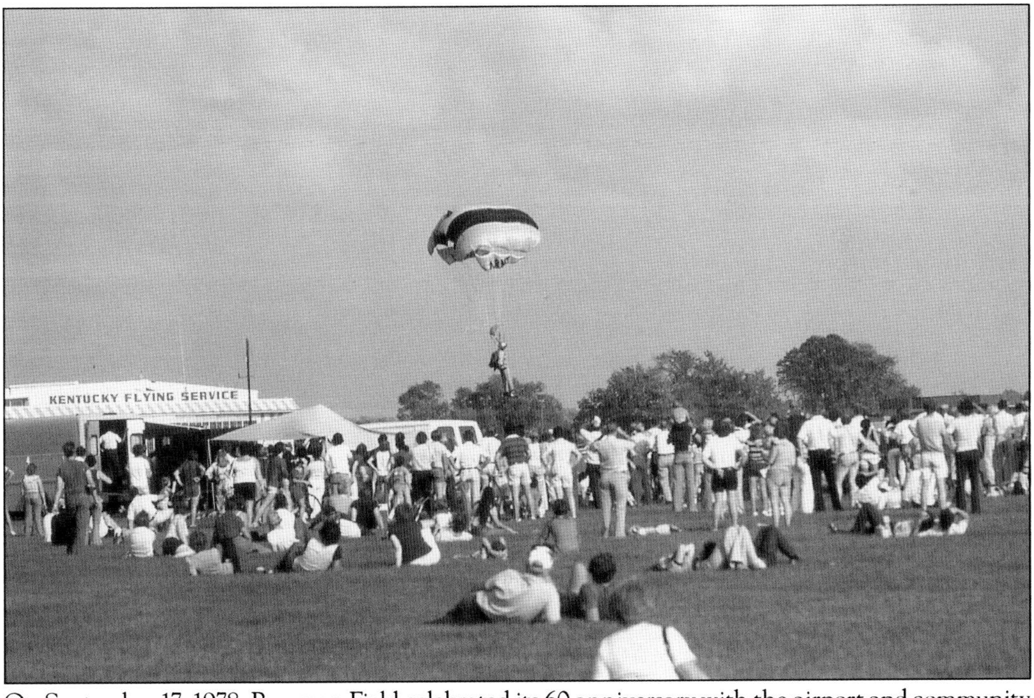

On September 17, 1978, Bowman Field celebrated its 60 anniversary with the airport and community coming together to put on an event. Since this was on a Sunday afternoon, many came to the airport to enjoy activities led by a parachute jump by the US Army's Golden Knights parachute team. Airplane and helicopter rides were offered, and music was provided by the Bowman-based 100th Army Division Band. (Author's collection.)

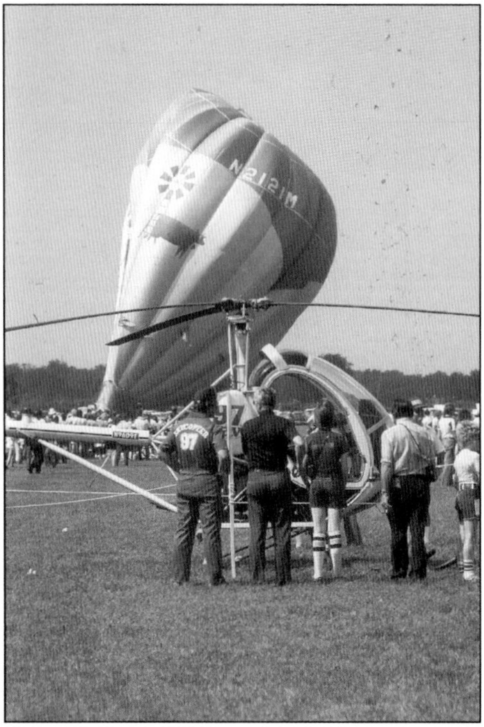

The 60th anniversary celebration at Bowman Field had something for everyone, including a hot-air balloon race. Lifting off among admirers, this Ehrler's Dairy balloon heads skyward. Ehrler's Dairy was a pioneer in Louisville balloon racing. The now-gone chain of dairy stores was famous for its delicious ice cream and candies. Tethered balloon rides were also available. (Author's collection.)

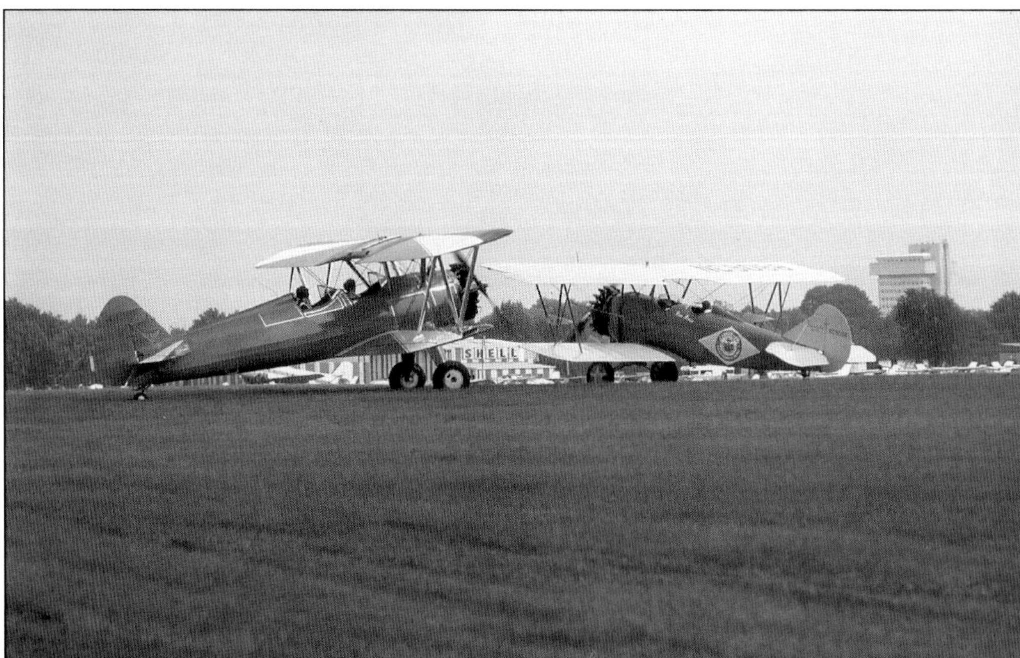

Of course, the public attends airport events to see airplanes, and the 60th anniversary of Bowman Field had some good ones. During the flying portion of the anniversary, a vintage Boeing Stearman passes a 1928 American Airlines TravelAir while taxiing across the airport. Within minutes, they would be in the air, with those on the ground following. Biplanes are always popular because they bring back memories of the early days of aviation. (Author's collection.)

Bowman Field's longest-serving organization is the Aero Club, which dates to 1922. In 1980, the club initiated an annual Bowman Air Fair that continued for a number of years. Each air fair spotlighted the best of Louisville's aviation community, with locals performing aerobatic routines, parachute jumps, and fly-bys. It also had an outstanding aerial demonstration that was performed by one of the nation's leading aerobatic pilots. This is the cover for the 1981 Bowman Air Fair program. (Author's collection.)

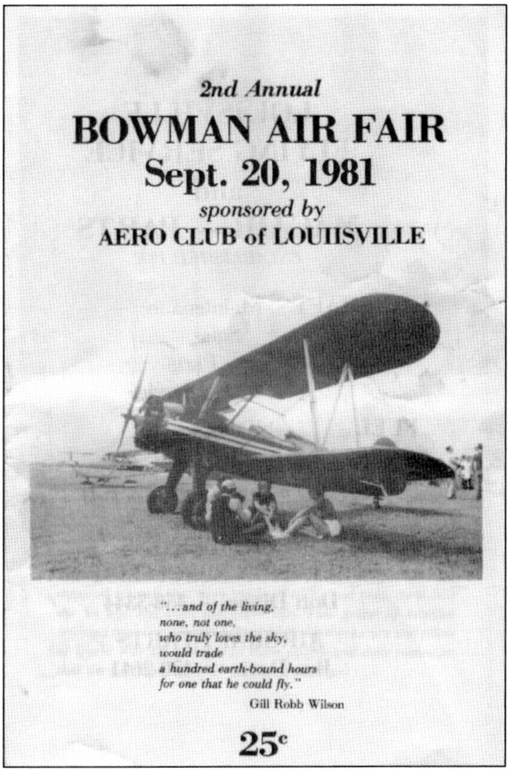

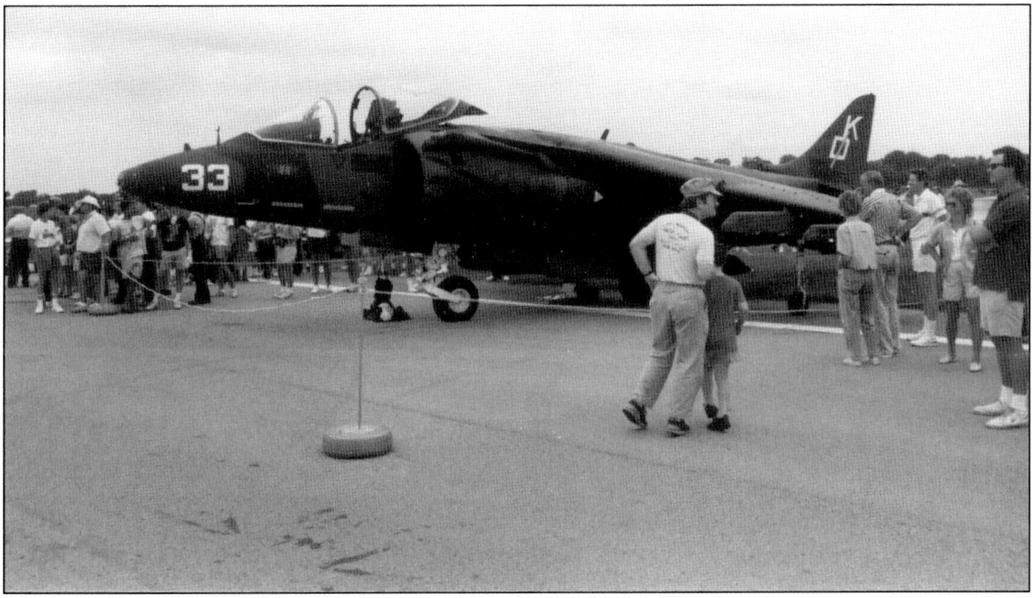

Over the years, the Bowman Air Fair displayed many aircraft both old and new. But it was the 1988 air fair that may have had the most unique airplane of them all. With spectators gathering around it, this US Marine Corps AV-8B Harrier jet could possibly be the fastest airplane ever to land at Bowman Field—at least at a public event. With a near-supersonic top speed, the Harrier is able to make vertical takeoffs and landings, which is the reason this high-performance jet could come to Bowman Field. (Author's collection.)

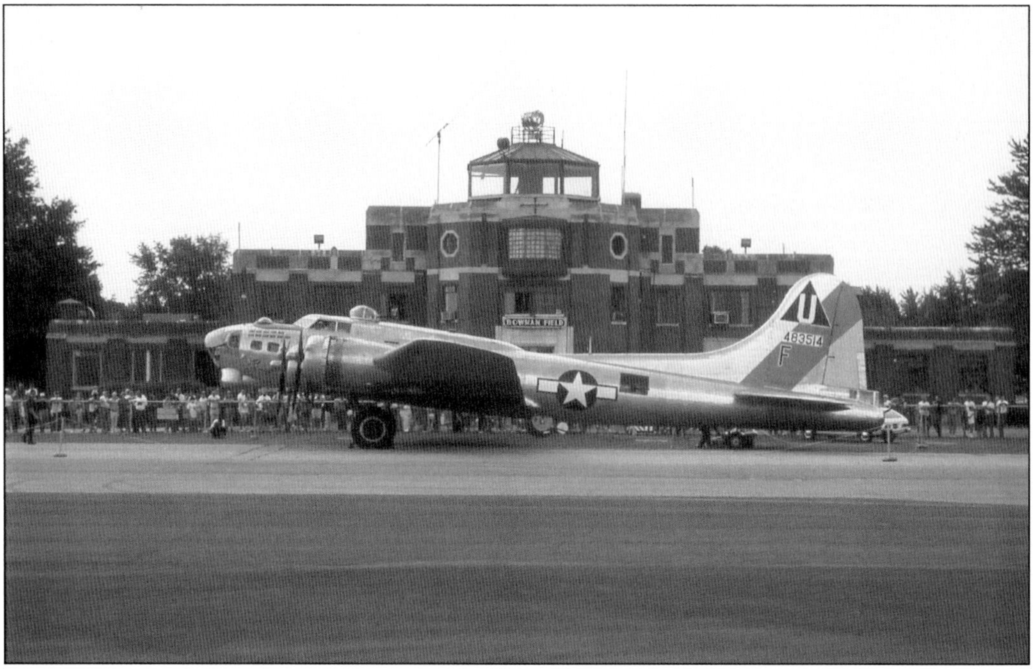

Historical aircraft have always been popular at Bowman Field, as seen in these photographs. Several noted organizations including the Experimental Aircraft Association and the Collins Foundation have historical World War II military aircraft that tour the country each year. Above, a B-17 Flying Fortress named "Sentimental Journey" from the Commemorative Air Force visits Bowman Field in 1987. Below is a B-24 Liberator from the Collins Foundation on view in 1989. In general, every year, a historical military aircraft or airliner will come to Bowman Field. (Both, author's collection.)

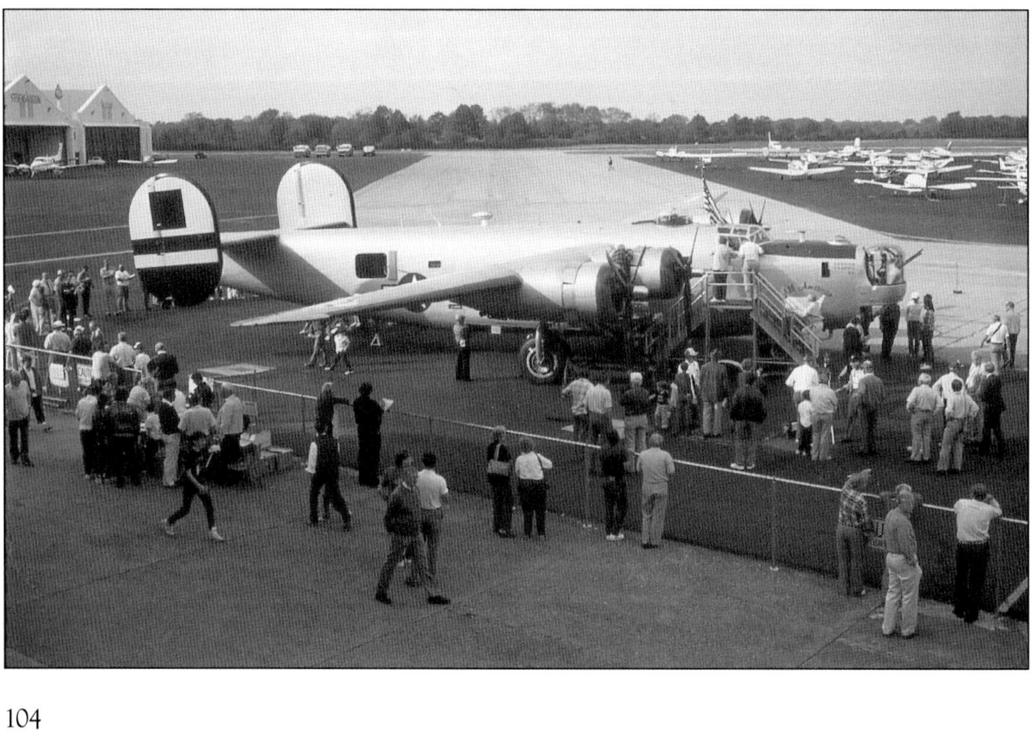

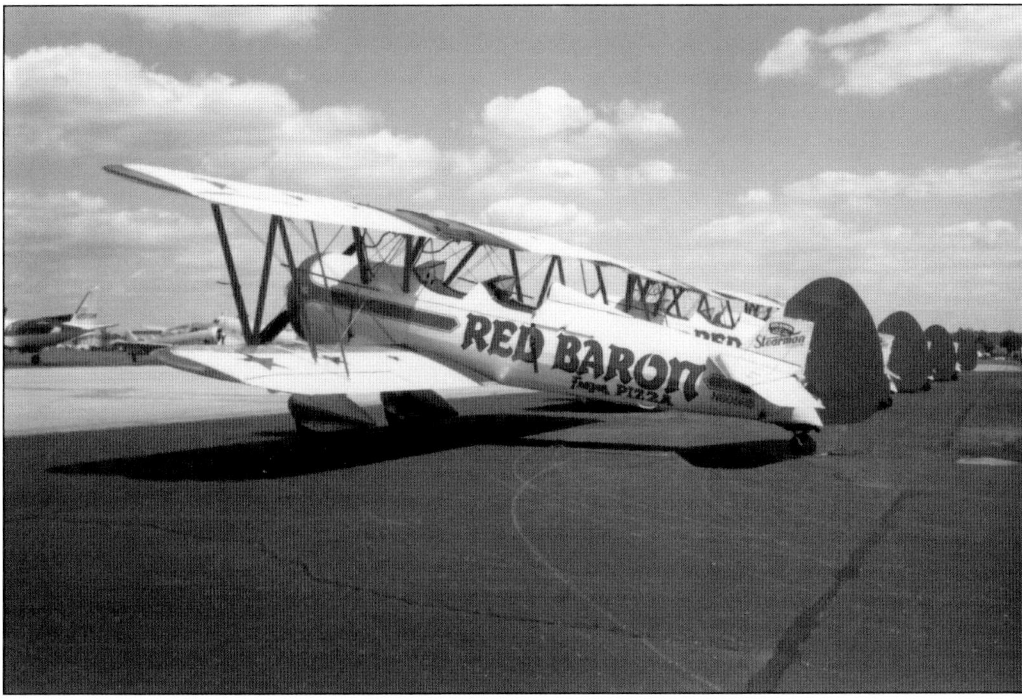

Nearly every year during the 1980s and 1990s, the Red Baron Pizza Squadron would visit Bowman Field. It was always popular, and the colorful red and white biplanes attracted attention. The team dates to 1979 and usually traveled with five Stearman biplanes. Often, they would give rides to employees of stores that sold Red Baron pizzas. The team was disbanded when the corporation that owned Red Baron Pizza was sold. (Both, author's collection.)

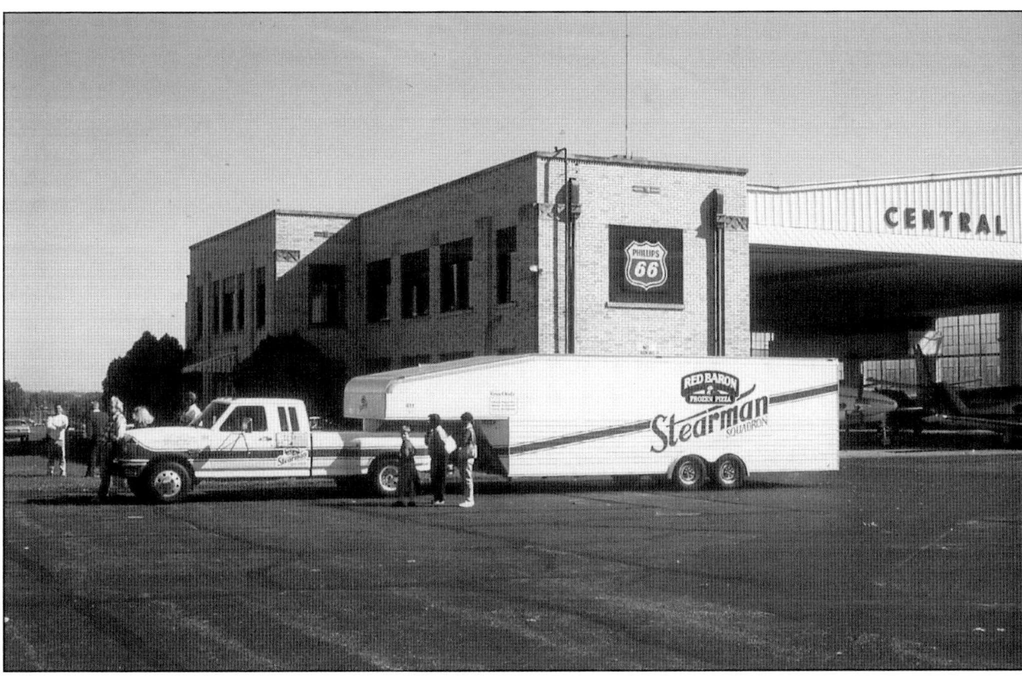

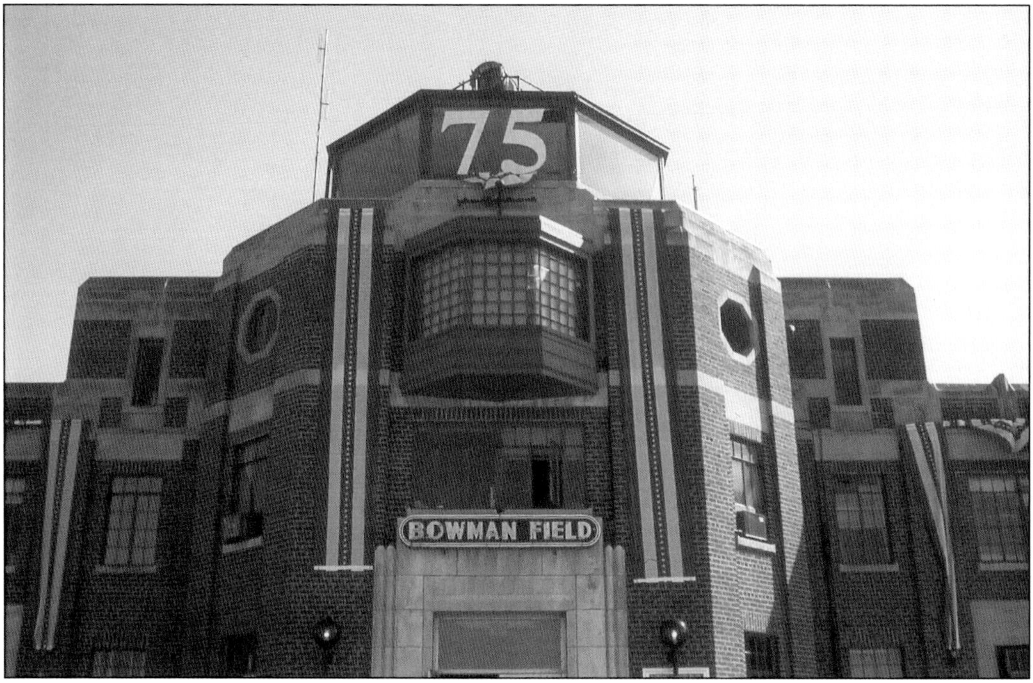

Bowman Field's 75th anniversary was celebrated on Saturday, April 23, 1994, which was the same day as the popular annual Thunder Over Louisville air show and fireworks event at the Ohio River. With two aviation events in one day, the city of Louisville was ready to look at airplanes, and it would not be disappointed. For Bowman Field, this was one of the largest events ever held at the airport. Visitors were able to walk the entire distance from the big east side hangar to the twin hangars on the west side. The administration building dazzled with decorations while a C-130 from the Kentucky Air National Guard and the Goodyear Blimp dominated the flight line. (Both, author's collection.)

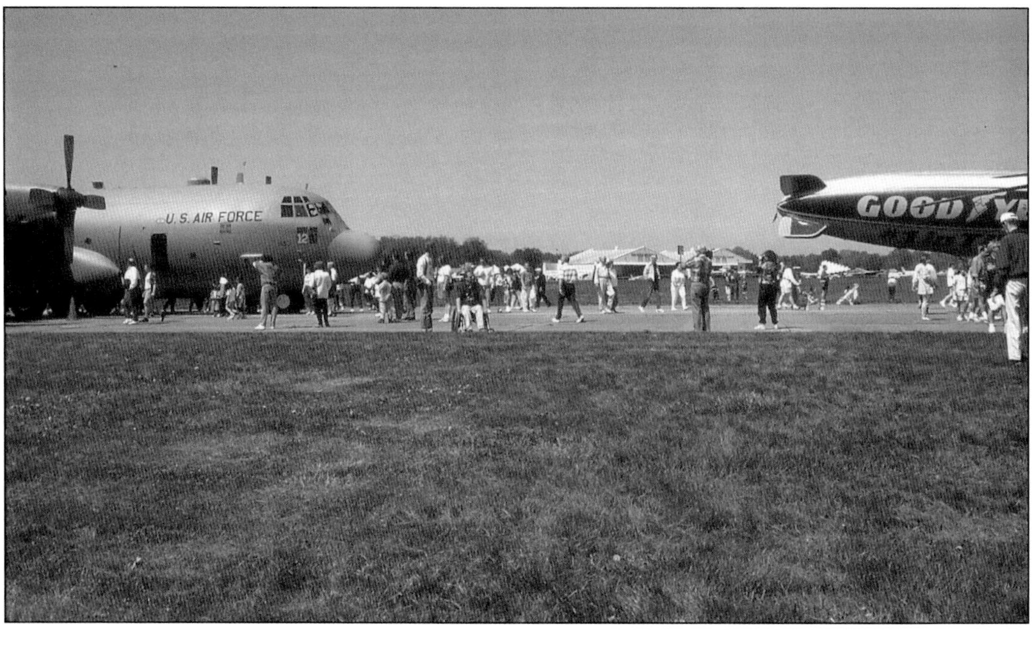

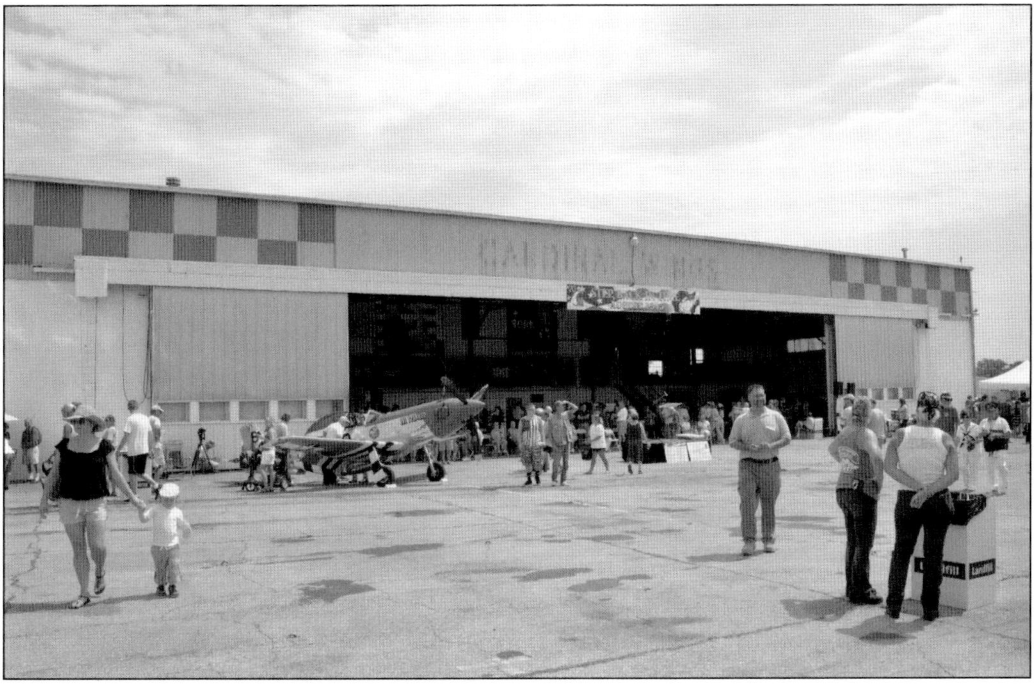

August 2015 marked the 70th anniversary of the end of World War II. To commemorate this, a "Spirit of 45" event was held on the east side of Bowman Field on August 15. Attendance was far greater than planned. Veterans were encouraged to attend, and participants were treated to displays of military equipment, warbirds, and reenactors. So successful was the 2015 Spirit of 45, it was decided to hold the event for the next several years. (Both, author's collection.)

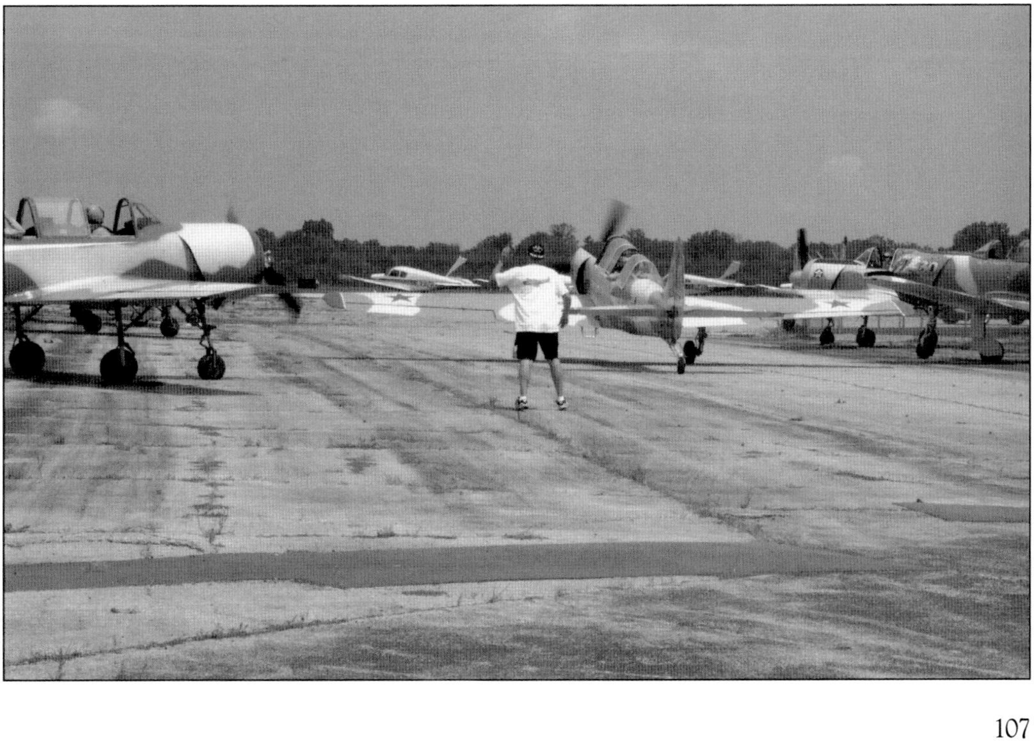

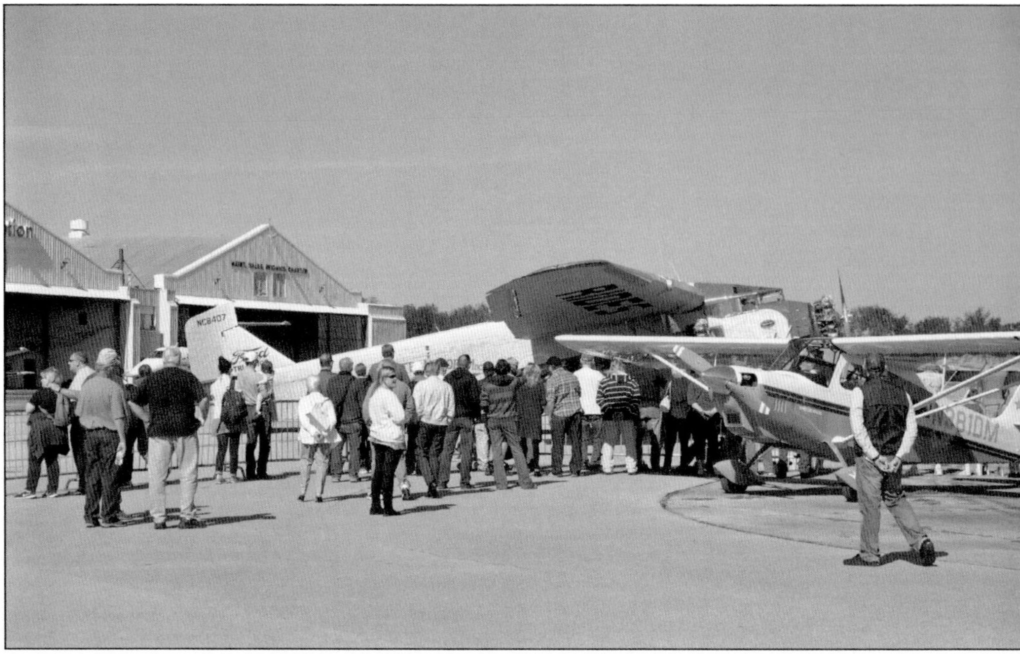

The Bowman Field Aviation Heritage Festival had a modest start in 2015, but hit its stride the next year. It featured displays of historical aircraft, with rides offered in several of them. Above, festival visitors are waiting for a ride in a beautifully restored Ford Tri-Motor. Another attention grabber was the Lockheed Model 12A Electra Jr from Georgia seen below. This same airplane first came to Bowman Field in 1937 or 1938 for the Kentucky Derby. (Both, author's collection.)

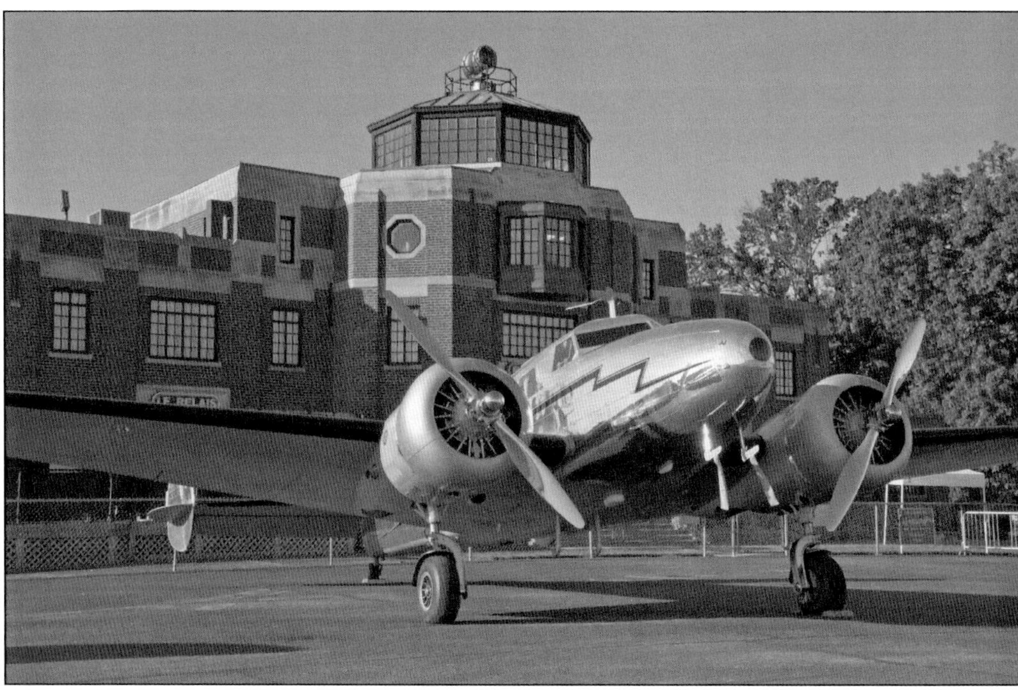

Six

LOOKING BACK, LOOKING FORWARD

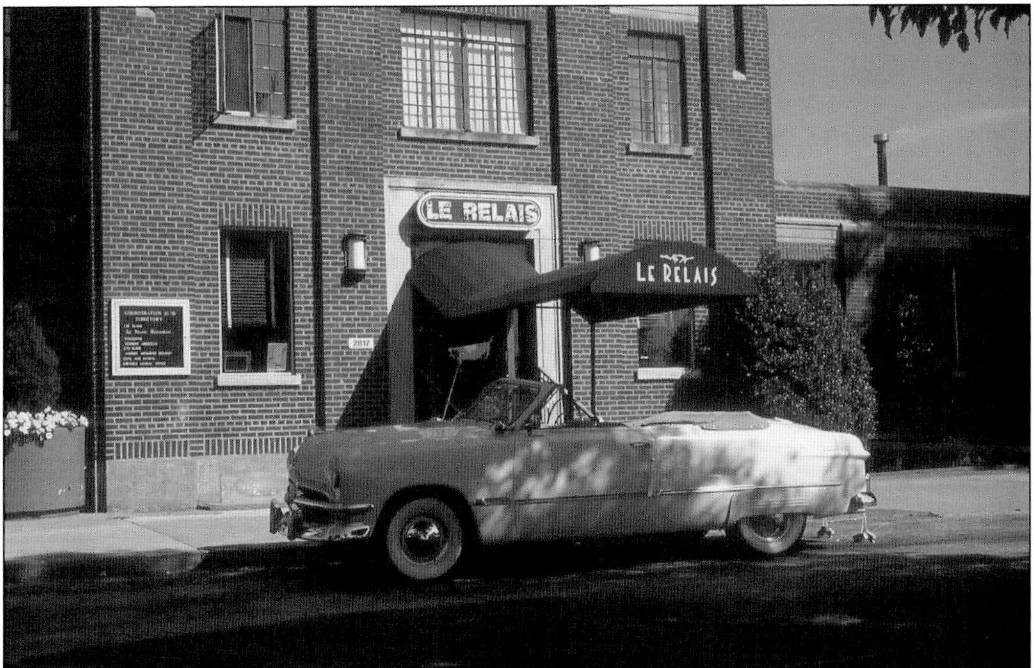

The Bowman Field administration building has had a restaurant for many years. Since the building was enlarged in 1936, restaurants have come and gone. They include Harry Ackley's, American Airlines Sky Chief, Bill Boland's, Jimmie Lynn's, Hoe Kow, Bico Mattei, and Le Relais. Le Relais has been the longest serving restaurant, opening in 1988. All have been located in the east wing of the building, and to add space, Le Relais added an outdoor deck that faces the airfield. (Author's collection.)

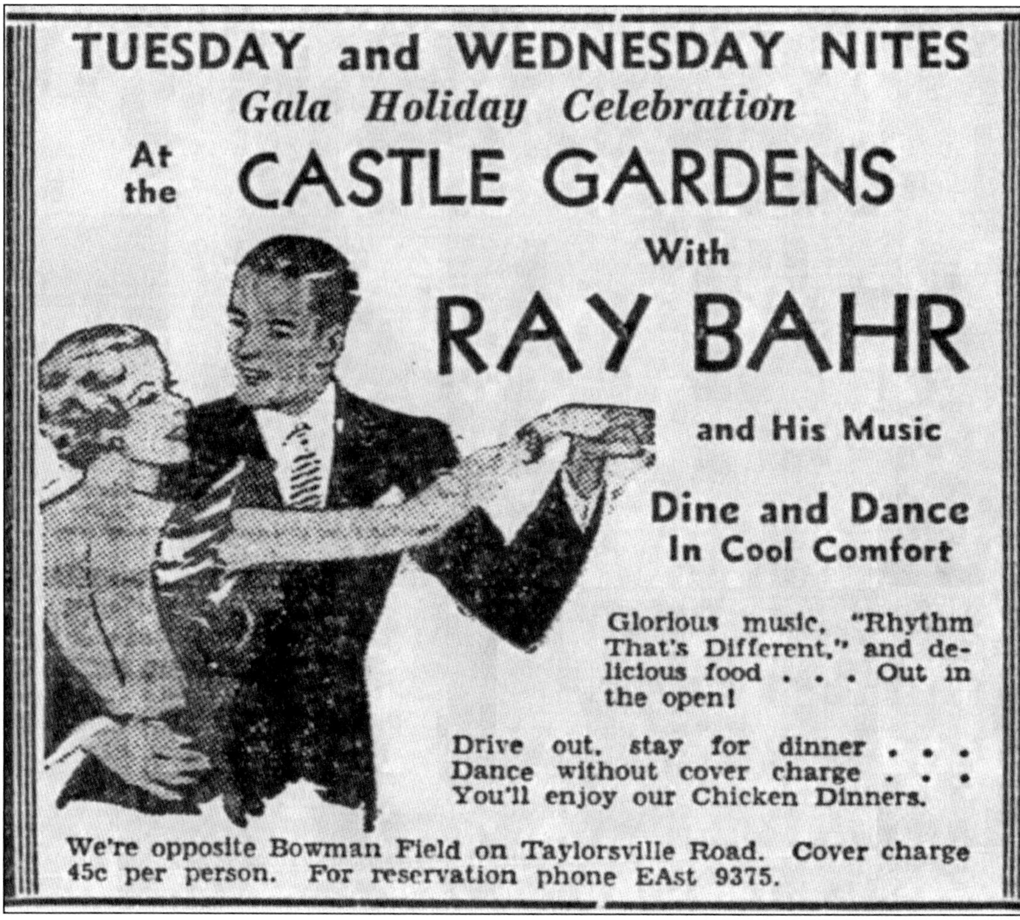

Air Devils Inn (ADI), located across Taylorsville Road from the administration building, has long been a watering hole for Bowman Field aviators. Although it has had the Air Devils Inn name since 1934, the building was originally the Maple Grove School. Switching to entertainment, the location was first known as Castle Gardens, and at that time, it heavily advertised "Ray Bahr," as its house band. During its early years, ADI featured outdoor dancing and had a large following. Currently, it is an indoor venue. (Both, author's collection.)

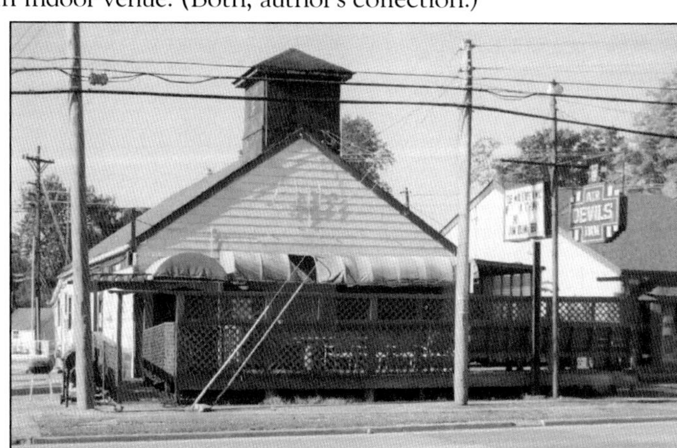

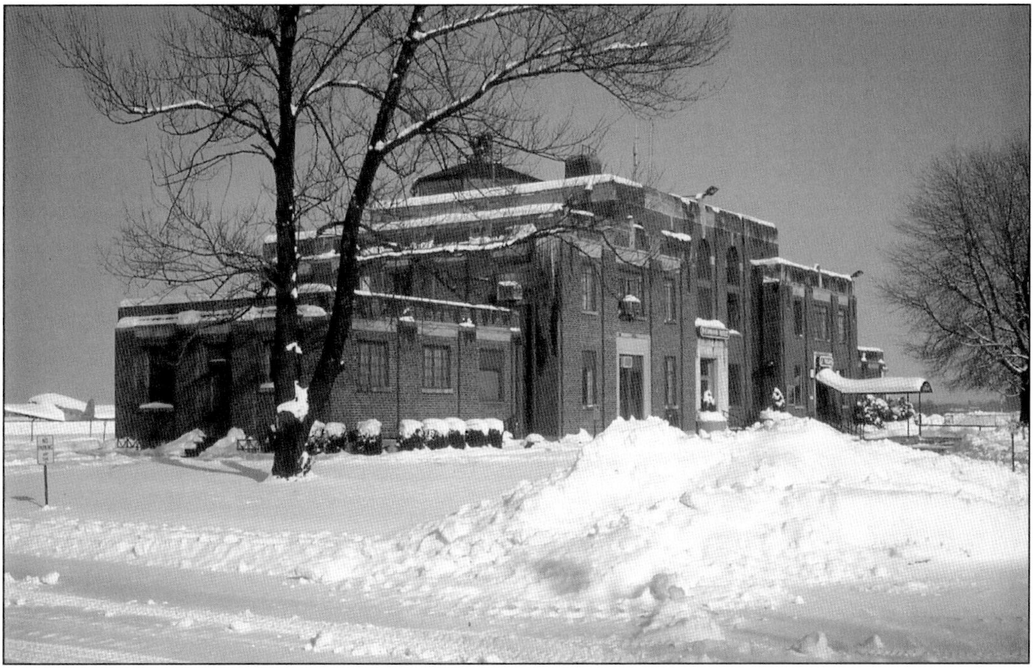

The one building that defines historic Bowman Field is the administration building. First constructed in 1929 and enlarged in 1936, this building has been well maintained and is still currently in use. A variety of businesses have offices inside, and its large center lobby is dedicated to the history of the airport with photographs and displays. Over the decades, the building has endured the seasons and been decorated for the holidays. It is open seven days a week and begs those passing by to stop and take a look. (Both, author's collection.)

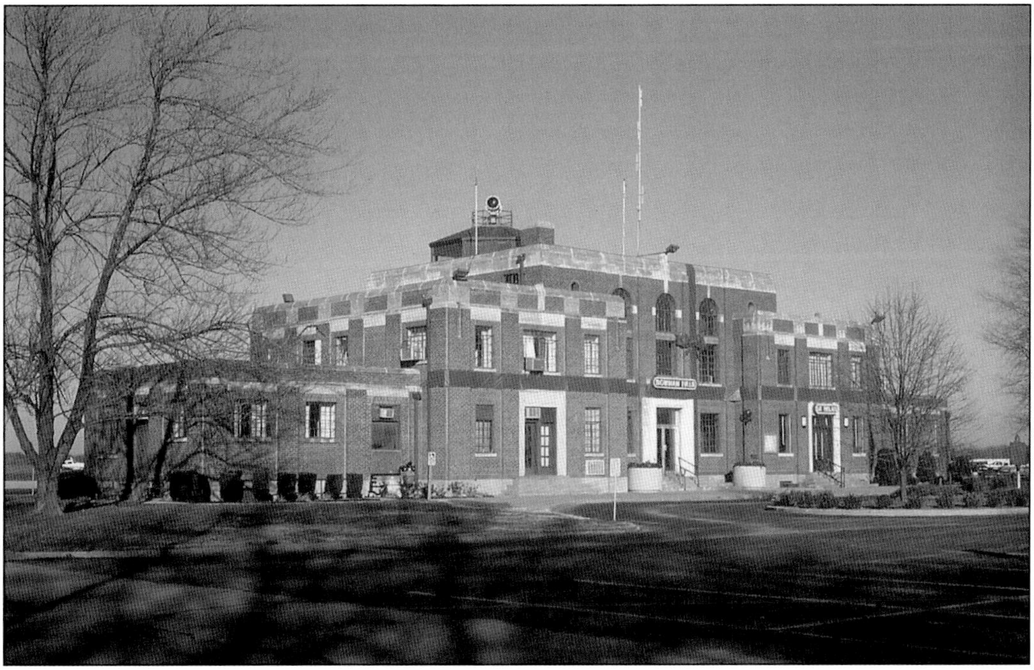

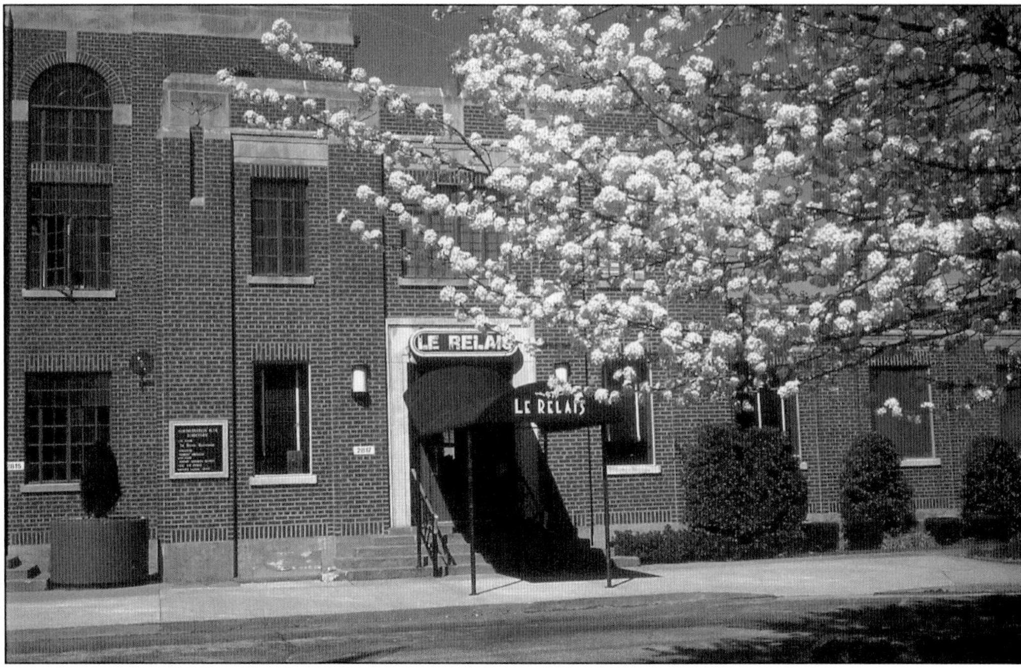

The Art Deco design of the Bowman Field administration building has stood the test of time and casts a charm upon all who see it. Highlighted by springtime color or the bare branches of fall, it rises toward the center and points to the sky—the domain of the airplane—seems to say, "This is the airport, this is Bowman Field. May this wonderful building, with its charm, look forward to a long bright future. (Both, author's collection.)

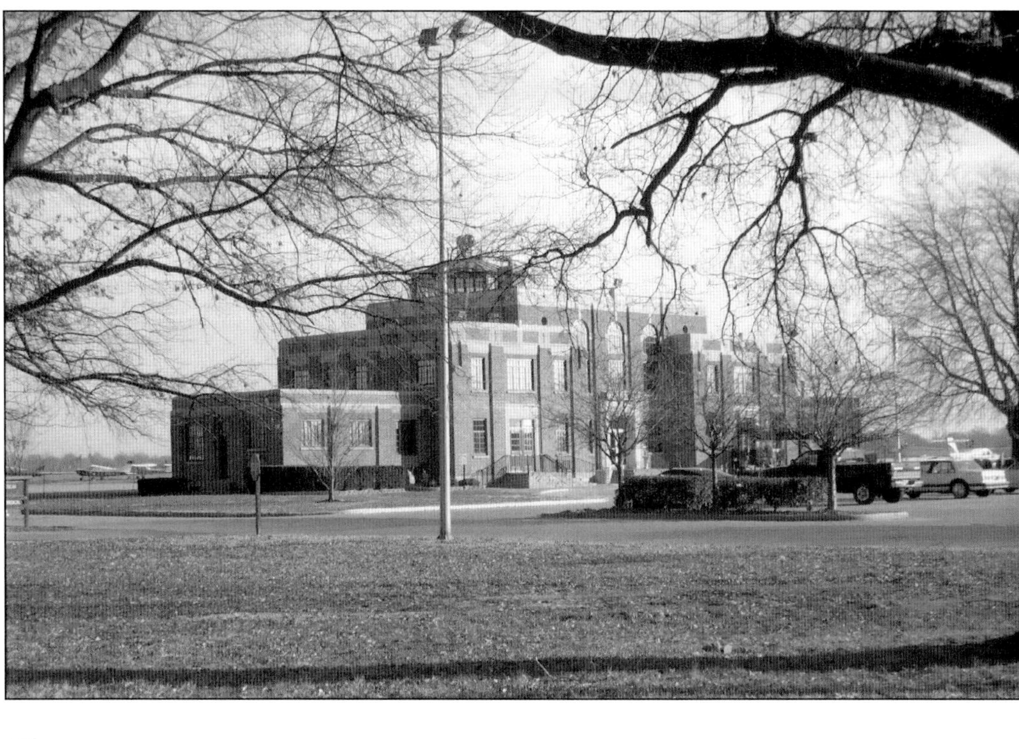

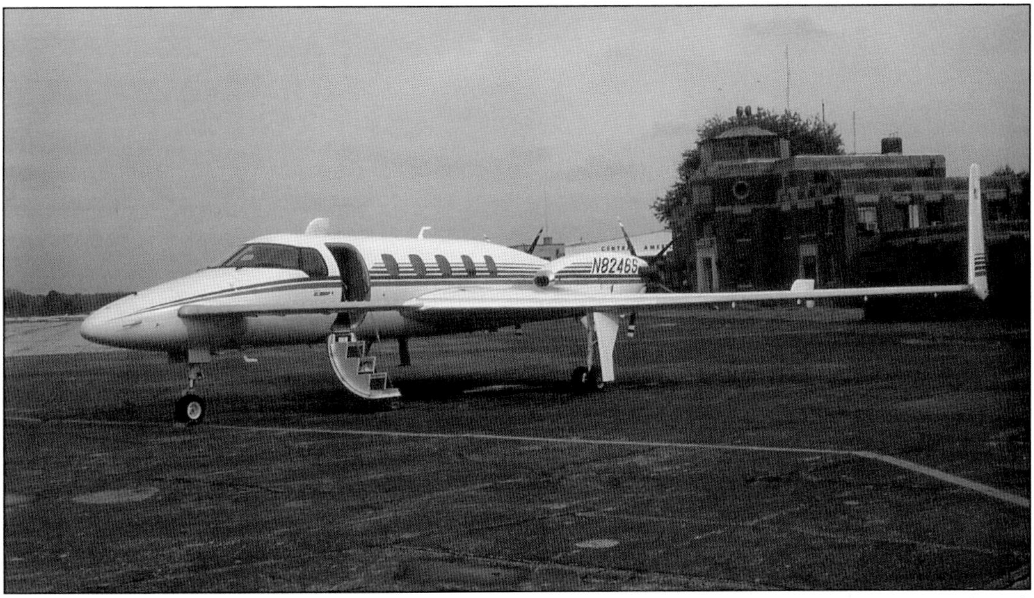

One of the most unusual airplanes seen at Bowman Field since World War II was the Beechcraft Starship. Development of the Starship dates to the 1980s, and it was an early user of composite materials. It had a futuristic appearance, with two rear-facing engines and a forward canard. Beechcraft made 53 Starships, sold a few, and leased many of the others. In 2003, the company terminated the project and scrapped the lease aircraft. A few are still flying, making it quite a rare airplane. This one is seen at Bowman Field in 1996. (Author's collection.)

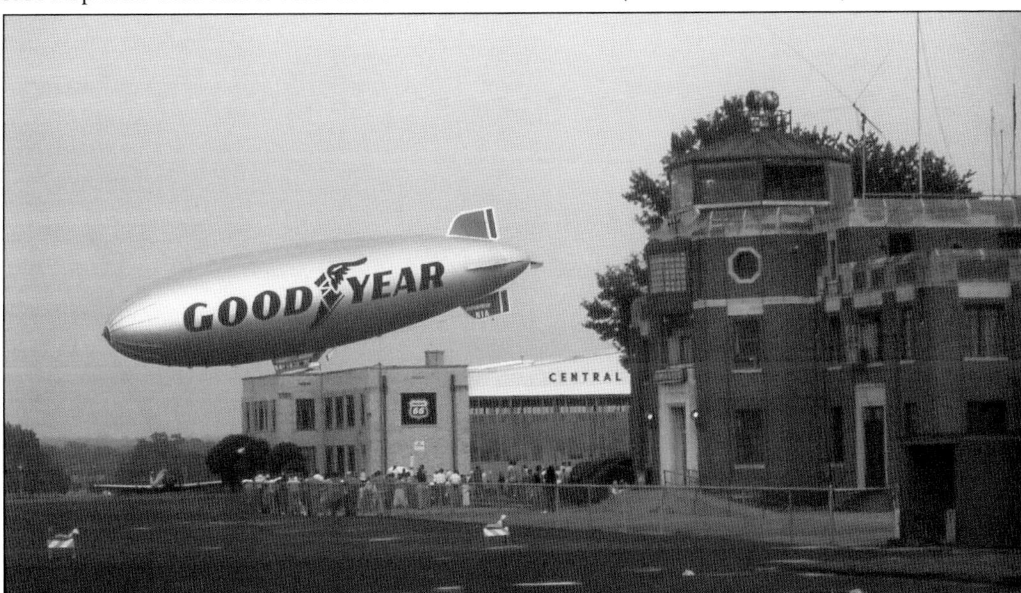

Several blimps visit Bowman Field each year. Of these, the Goodyear Blimp has been the most common, with the first one appearing at the field in the years before World War II. Blimps come to Louisville for various reasons, including as an aerial camera platform for the Kentucky Derby and other sports events. Others fly as aerial billboards to advertise sponsoring companies. (Author's collection.)

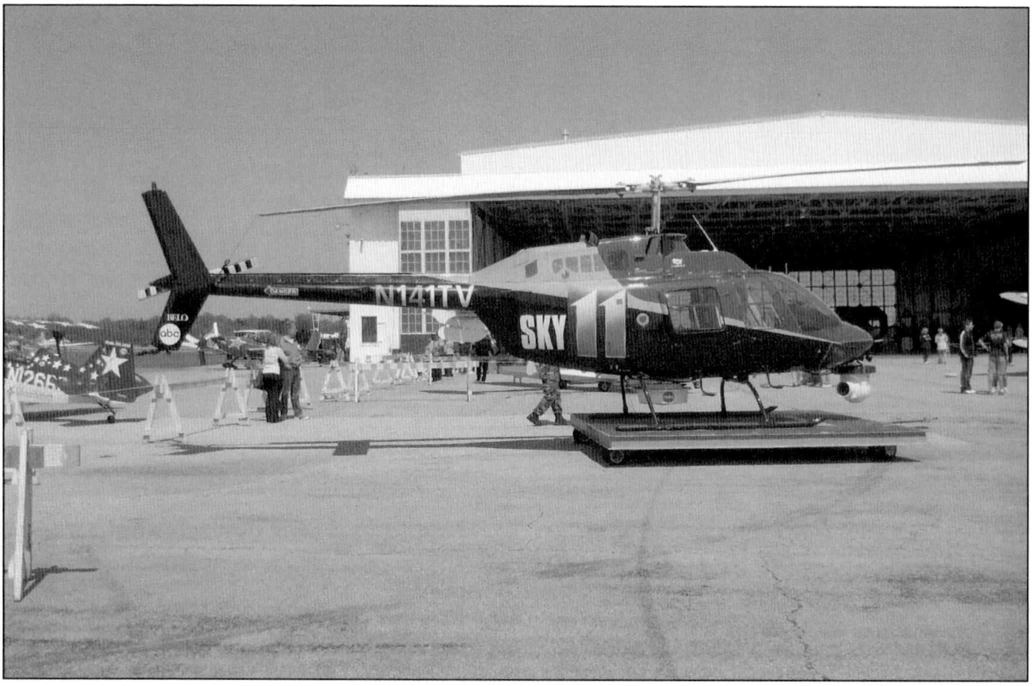

Helicopters have been seen at Bowman Field for a long time. The first recorded visit of a helicopter came in October 1944 when a US Army Sikorsky R-4 flew in for a demonstration to personnel at the Bowman Army Air Base. On a daily basis, the most commonly seen helicopters are the ones operated for television news. They include one jointly used by WAVE3 and WHAS11 and another one by WLKY32. The WLKY helicopter spends part of its time on the ground at Bowman Field and at the studio helipad on Mellwood Avenue. (Both, author's collection.)

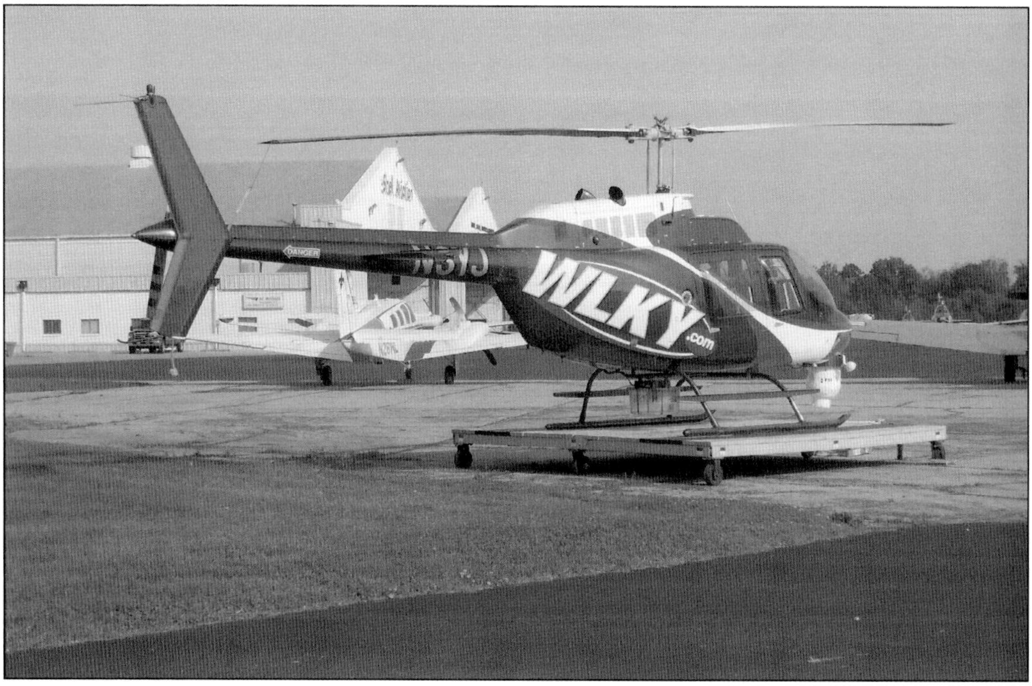

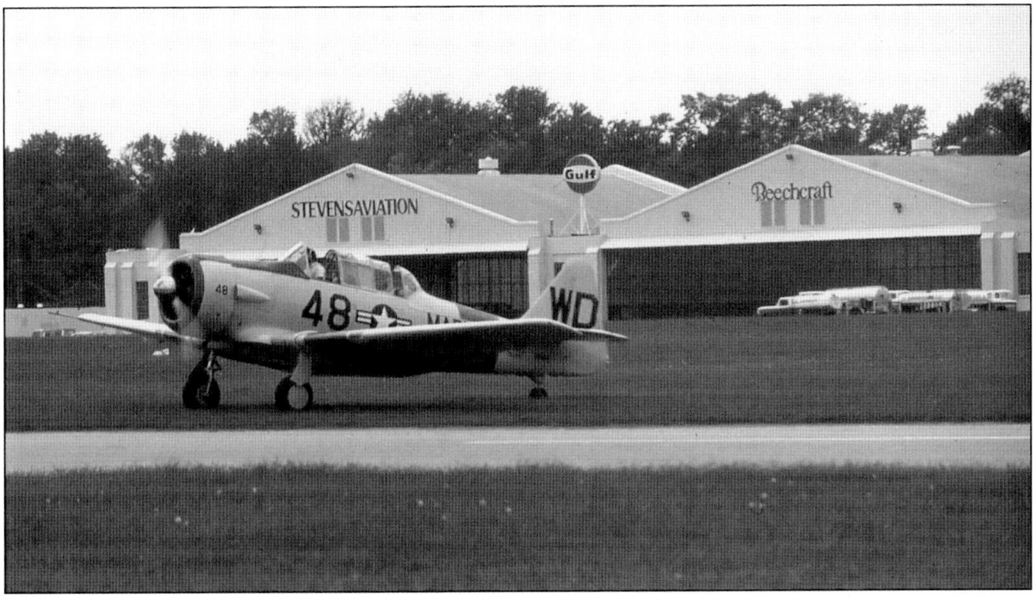

After being in business at Bowman Field for 50 years, Louisville Flying Service closed its doors in 1982. At this point it was LFS in name only, as original owners John C. Bennett and Frank Holzaphel had retired earlier. Taking the lease for the old twin hangars was Stevens Beechcraft, which was based in Greer, South Carolina. It established a flying service with 40 employees and provided charter operation along with aircraft sales and service. (Author's collection.)

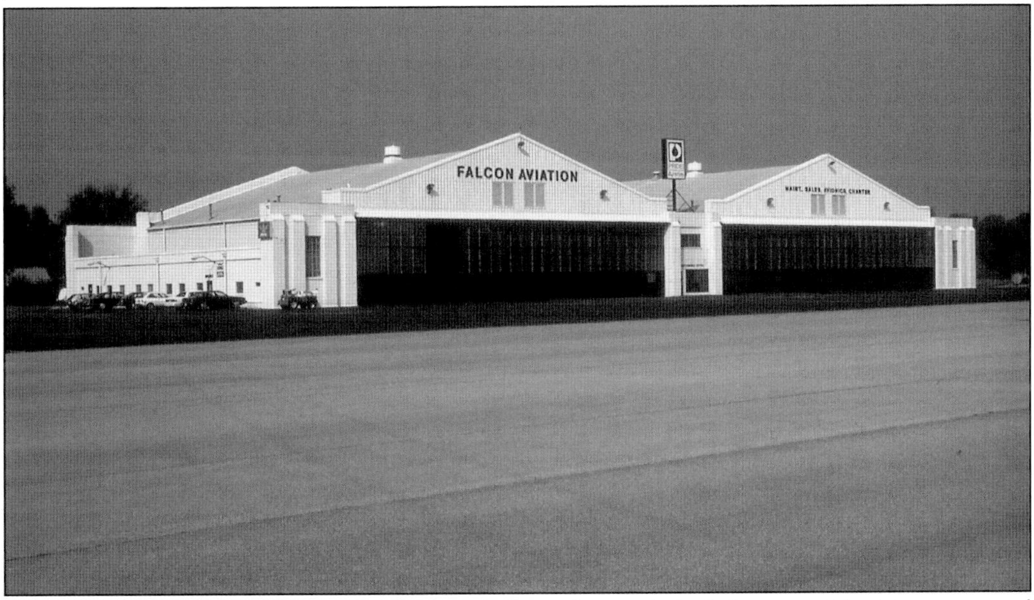

During the 1980s, general aviation was going through a period of change, with rising costs and less demand. Airport flying services suffered from the loss of business, which led to closures. Stevens Aviation at Bowman Field was one of those. Replacing Stevens in the twin hangar was Falcon Aviation owned by W.C. "Bill" Carroll. Carroll, originally employed by Stevens in 1977, started Falcon in 1981 and moved his business to the hangar in 1990. Falcon opened at a time when there was an increased demand for aircraft charter services, and Bill Carroll believed his business was well positioned to provide the service. (Author's collection.)

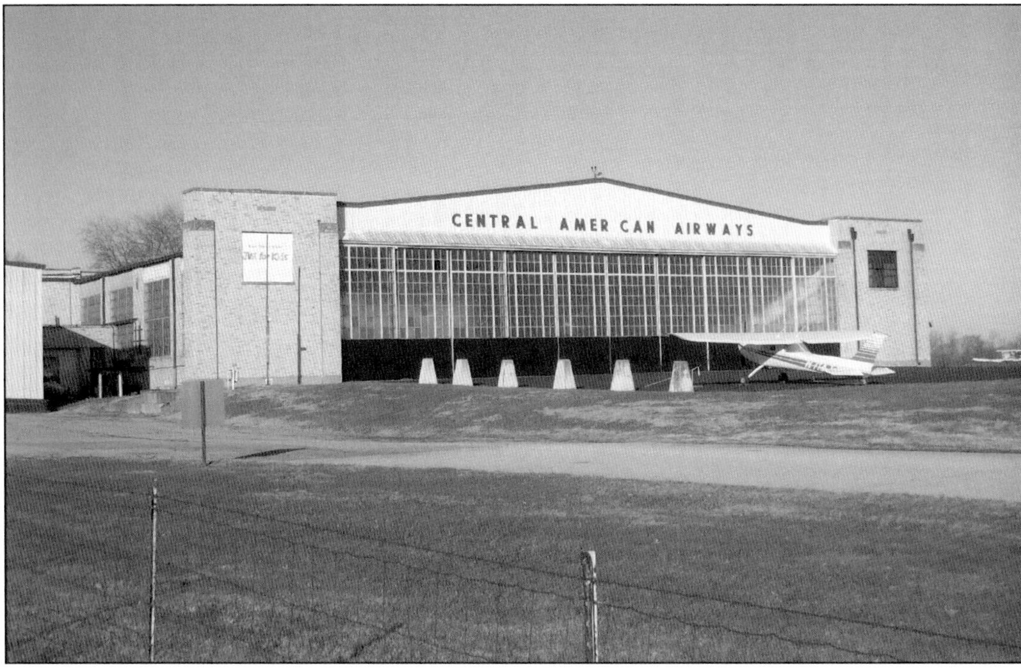

With over 70 years of service at Bowman Field, Central American Airways is now the field's oldest continuously operating flying service and is still family owned. It operates from two locations, on both sides of the administration building. In addition to its location in the Curtiss Flying Service hangar, which dates to 1950, it also uses the twin west-side hangars it acquired after Falcon Aviation closed during the early 1990s. CAA generates business from charter and servicing transient aircraft plus maintenance, storage, and service of hangared aircraft. (Both, author's collection.)

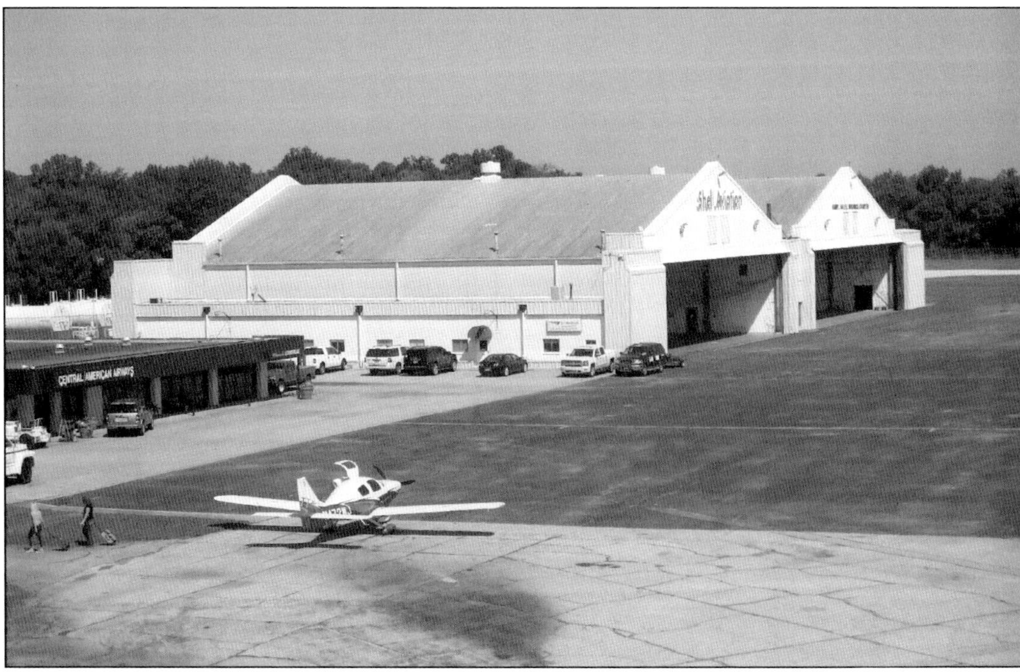

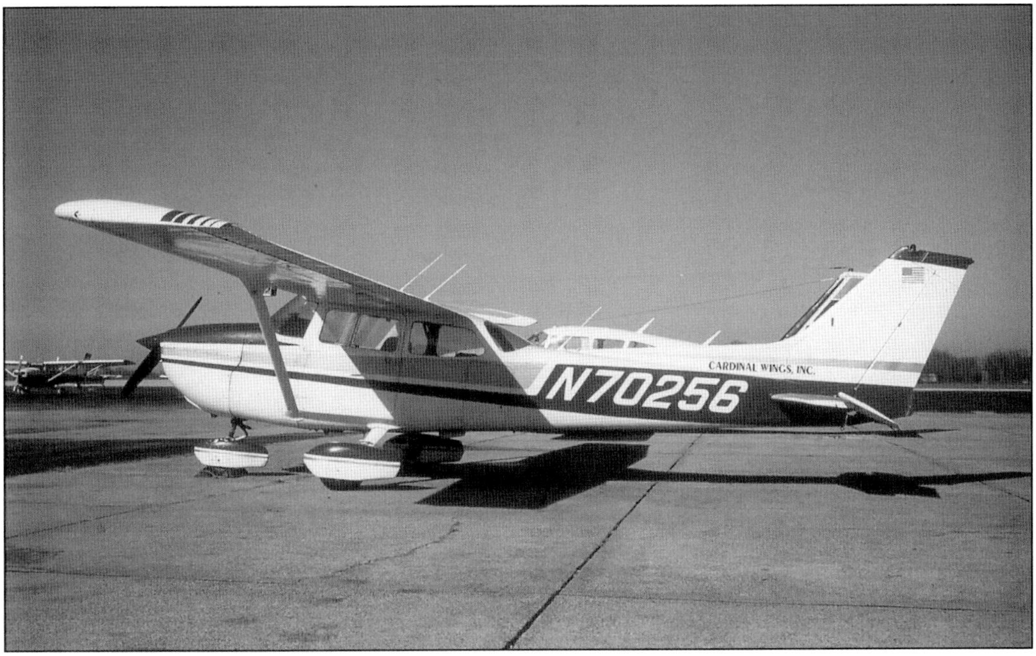

The Cardinal name has been well utilized at Bowman Field over the years. The latest is Cardinal Wings, which was started in the early 1990s spearheaded by Laura Benson. In five years, it was operating 16 airplanes. During its years of operation, it was located both in Hangar 5 and in the big east-side hangar. More recently, a new Cardinal Wings has emerged with pilot training, aircraft service, and sales. Although not associated with the new entity, Laura Benson Jones is still active at Bowman Field founding and supporting Flight Club 502, a flight training club for and run by high school students. (Both, author's collection.)

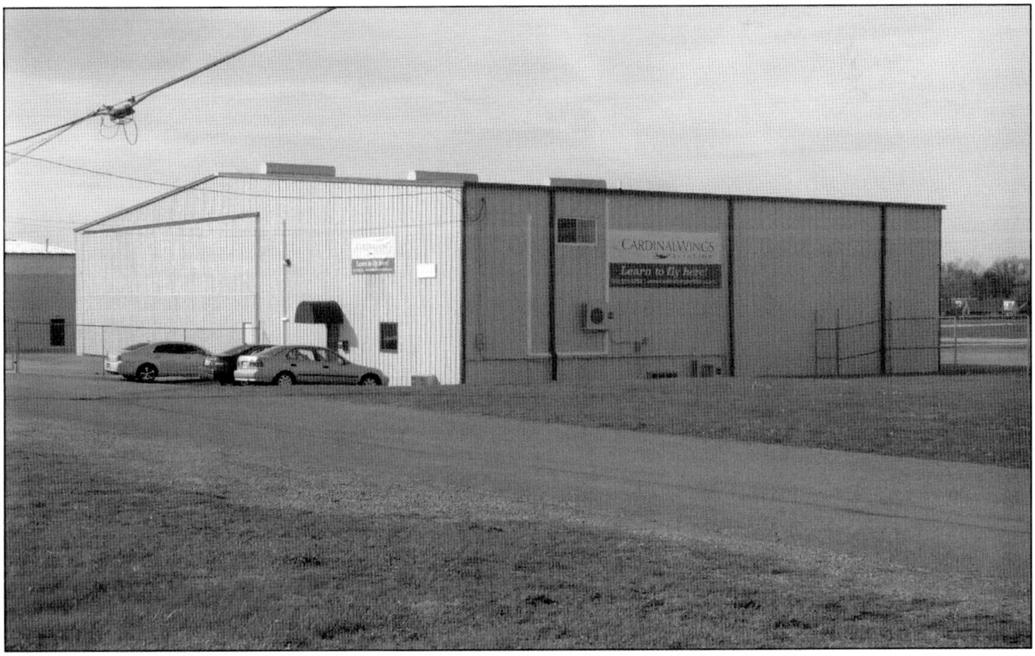

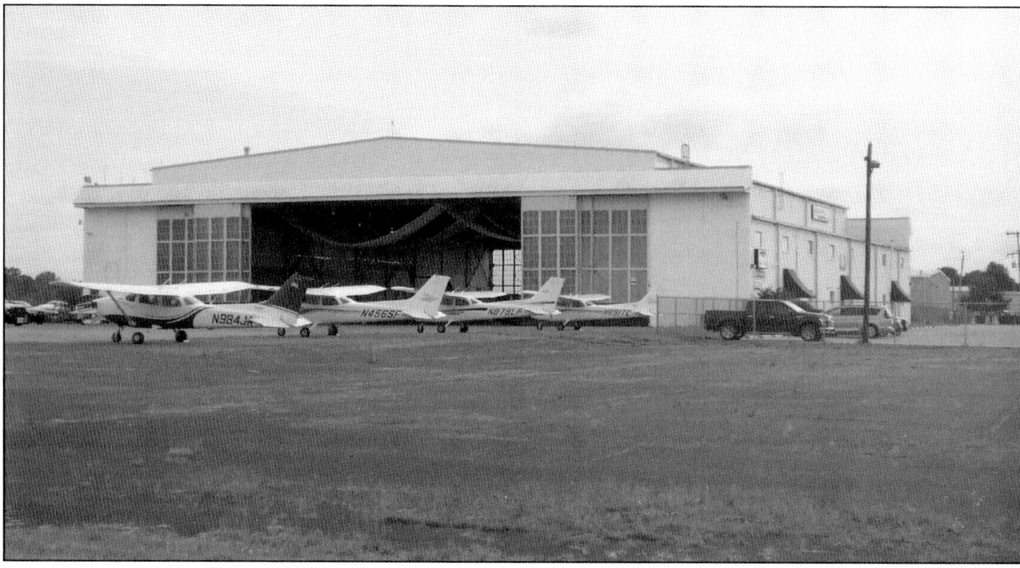

Succeeding Triangle Flying Service at Bowman Field on October 1, 2006, was Louisville Executive Aviation (LEA). Louisville businessman Ron Geary took the helm of LEA with the intention of modeling the service after Miami Executive Aviation (MEA) at Opa-Locka Airport in Florida. His partners were the owners of MEA. Believing in the future of secondary airports, Geary said, "We want to focus on the small airports and the small communities." In the years since, LEA has grown the largest general aviation service at Bowman Field and makes its hangar available for rental to weddings, receptions, meetings, and special events. It is still in operation in 2017. (Author's collection.)

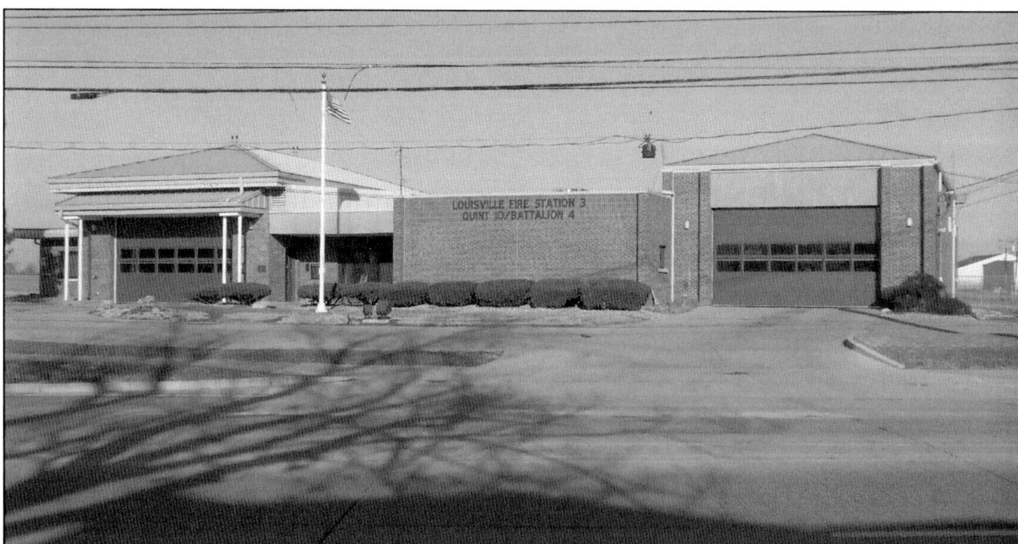

Airports are known to have accidents, and airplanes carrying large amounts of highly volatile gasoline can be a serious fire risk during a crash. This has made fire protection at Bowman Field important. During World War II, the Bowman Army Air Base was protected by a military fire department, while the postwar years have seen fire protection provided by Louisville Fire Station 3 at Taylorsville Road and Dutchmans Lane. This station, which belongs to the Louisville Fire Department and dates to 1956, serves both the airport and surrounding community. (Author's collection.)

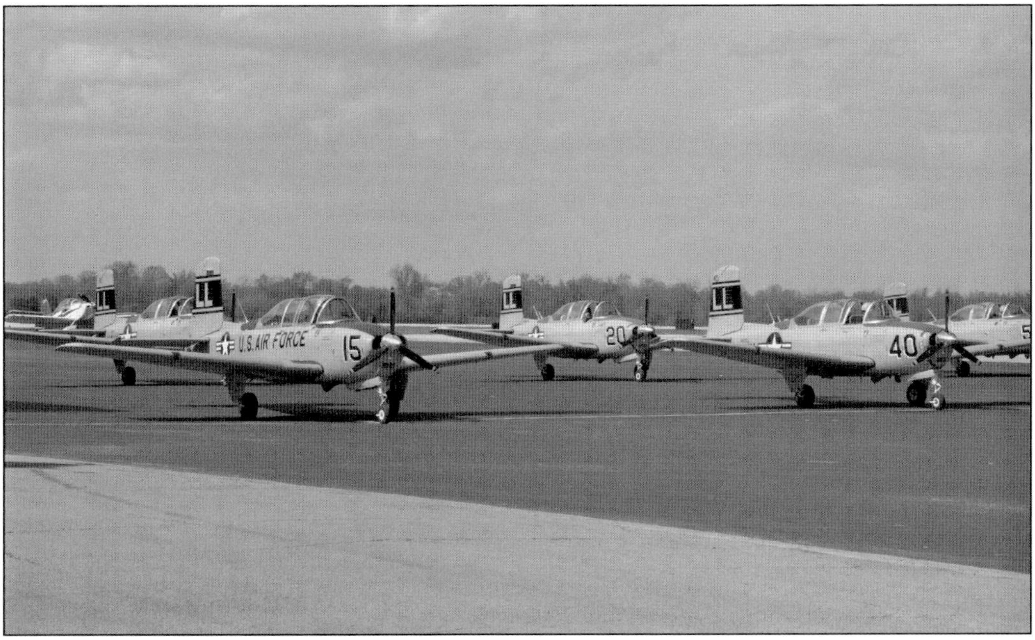

Since its inception in 1990, Thunder Over Louisville has been one of the most popular annual events that lead up to the Kentucky Derby. It is an air show and fireworks event over the Ohio River at Louisville's downtown waterfront. It is usually held two weeks before the derby and has both military and civilian aircraft for the air show portion. Generally, smaller civilian aircraft operate from Bowman Field, while military aircraft use Louisville International Airport. One of the longest participating flying groups is the Lima Lima team from Illinois. It flies several 1950 vintage Beech T-34s, as seen above. Another participant is the Trojan Horsemen team with big North American T-28s, pictured below. (Both, author's collection.)

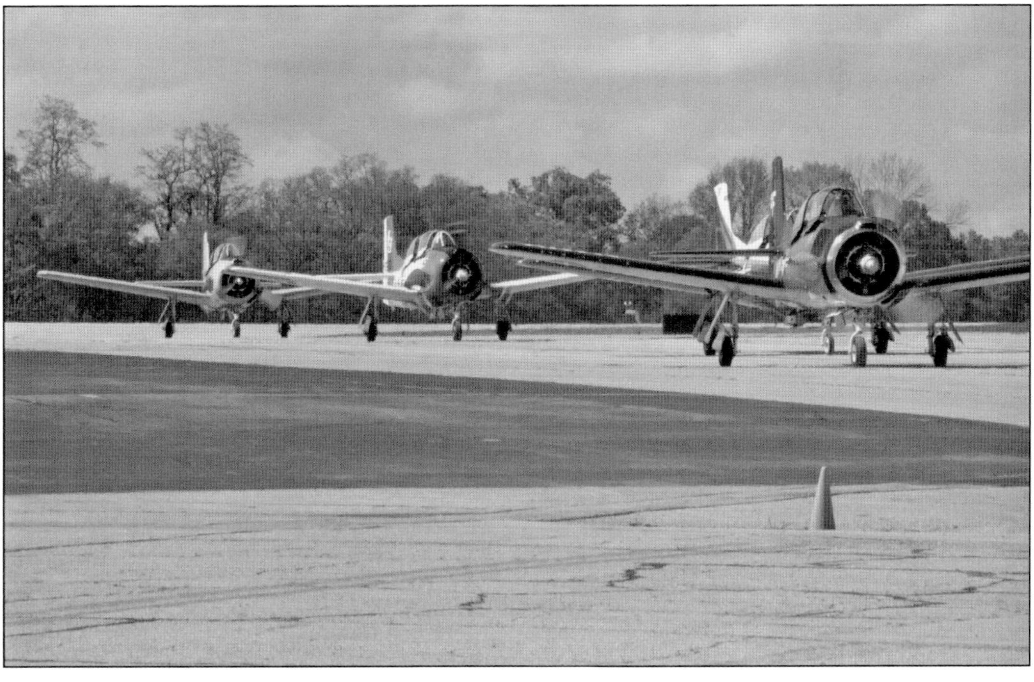

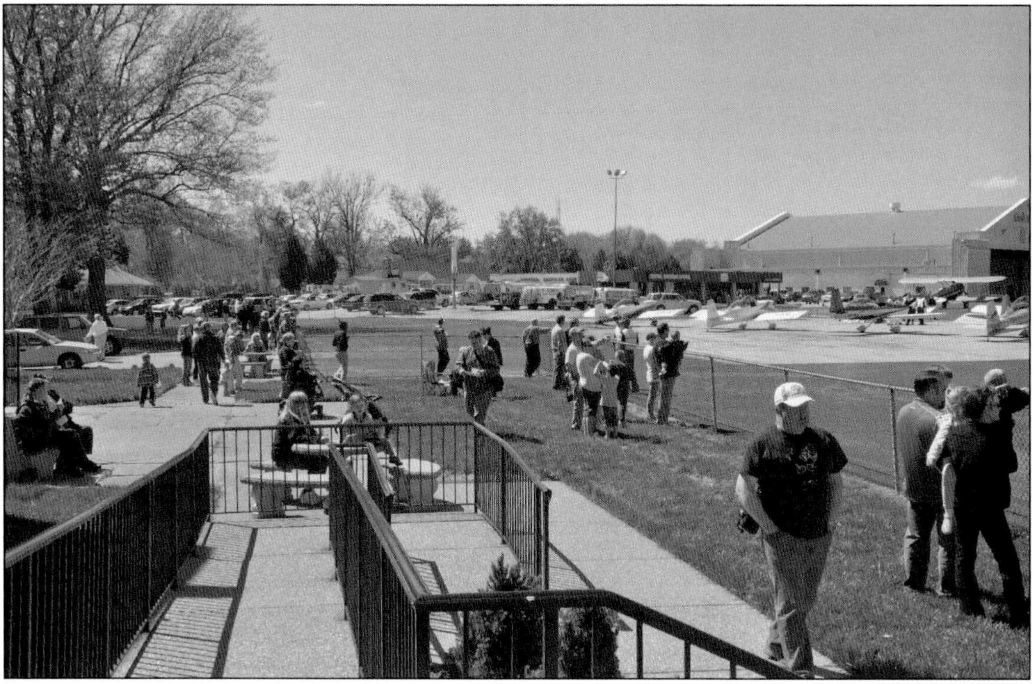

Thunder Over Louisville has always been a big event, with thousands gathering at the Ohio River to watch the air show and fireworks. The same holds true at Bowman Field. Entire families come to the airport to watch takeoffs and landings of aircraft and to enjoy the day. Of course, a comparison has to be made between this and the large gatherings of people at Bowman Field during the 1930s. If nothing else, it shows that airplanes are still magic and provide a sense of thrill and excitement. (Both, author's collection.)

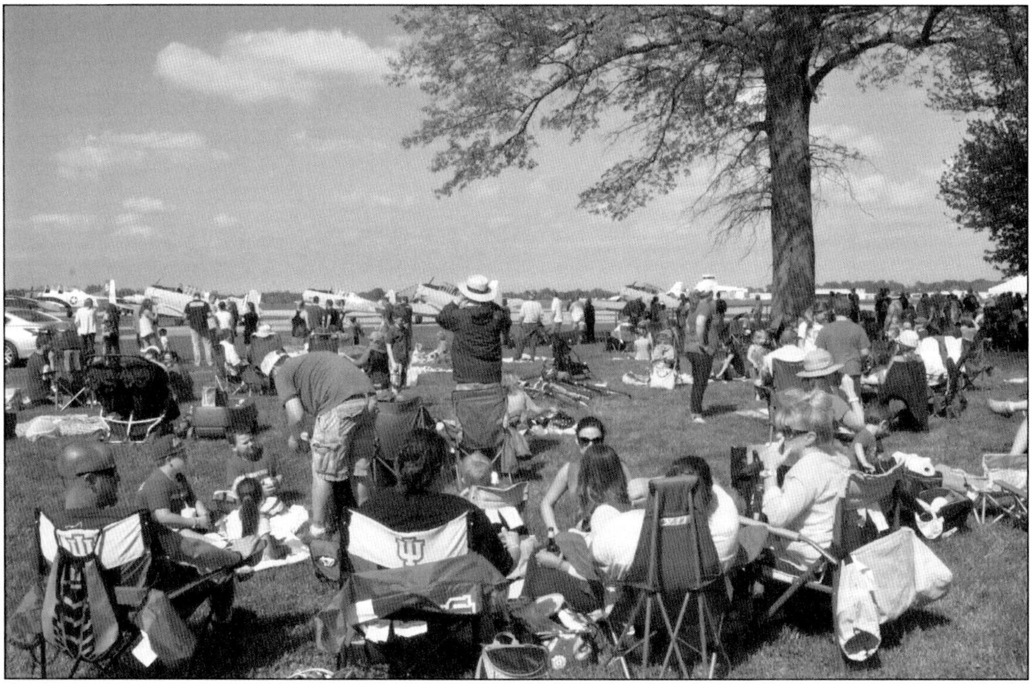

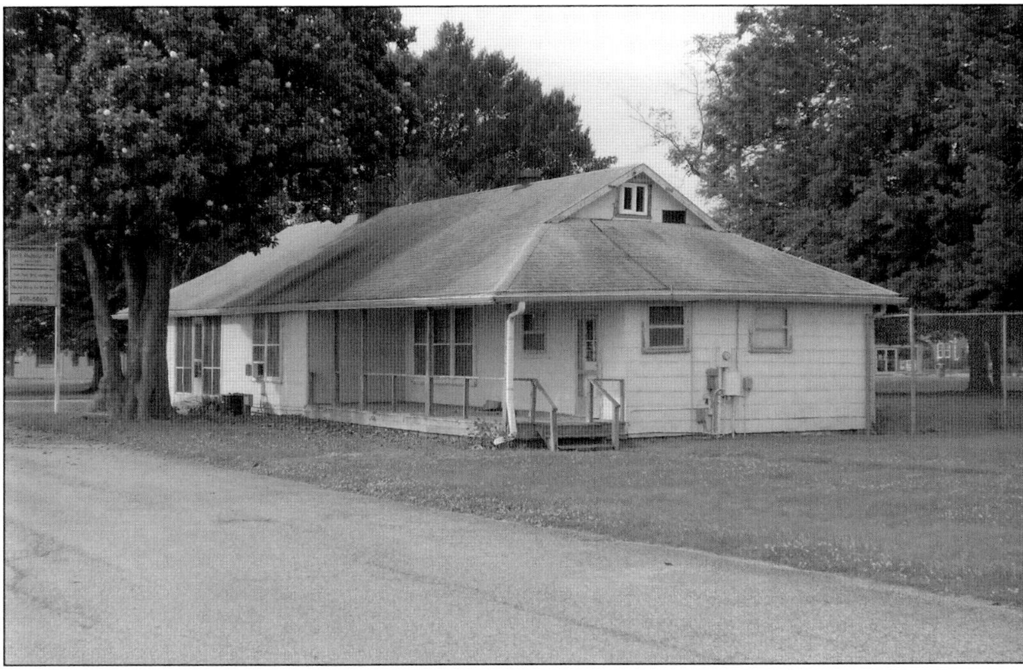

The oldest buildings at Bowman Field are the two west side cottages that date to 1922. They were moved along with two metal hangars from Godman Field, Camp Knox, for the 465th Pursuit Squadron Reserve. Originally three cottages were brought to Bowman Field; one was destroyed by a fire before World War II. The two remaining have been moved at least once and renovated over the years. After the military, they were used as housing for the Bowman Field airport superintendent and as an office for various businesses. As of 2017, the south cottage is being used by a doctor who is an FAA medical examiner. (Both, author's collection.)

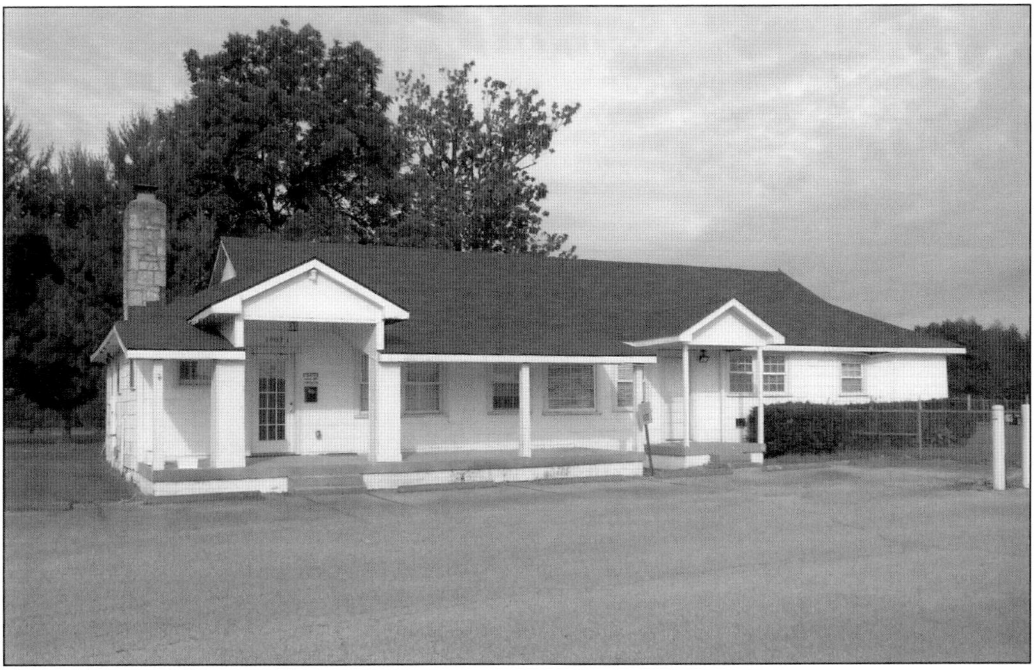

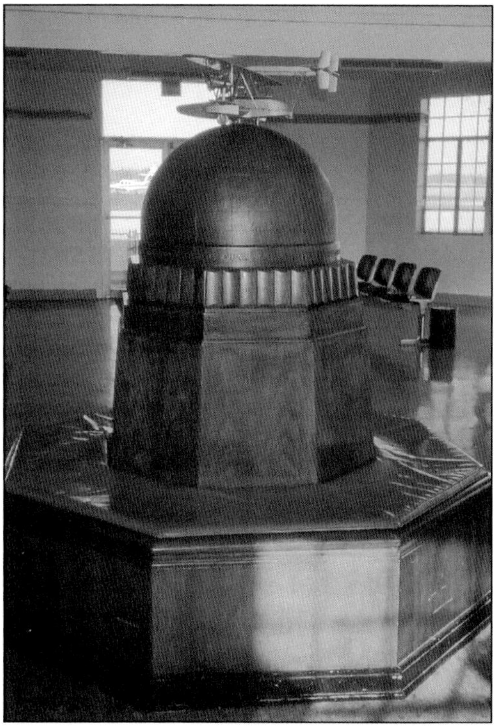

The Robert "Bob" Gast Memorial has been inside the Bowman Field administration building since 1937. It was made to honor Gast, who was one of the founders of Bowman Field in 1920, after he was killed in an aircraft accident in China in 1934. The memorial features a bronze hemisphere with a model of the Sikorsky S-38 airplane that Gast last flew on top and is mounted on an octagonal seat. Among those at the 1937 dedication were Gast's sister Mrs. Clifford Lusky, good friend William Curran, and Abram Bowman. The memorial was rededicated during a ceremony in 2015. (Author's collection.)

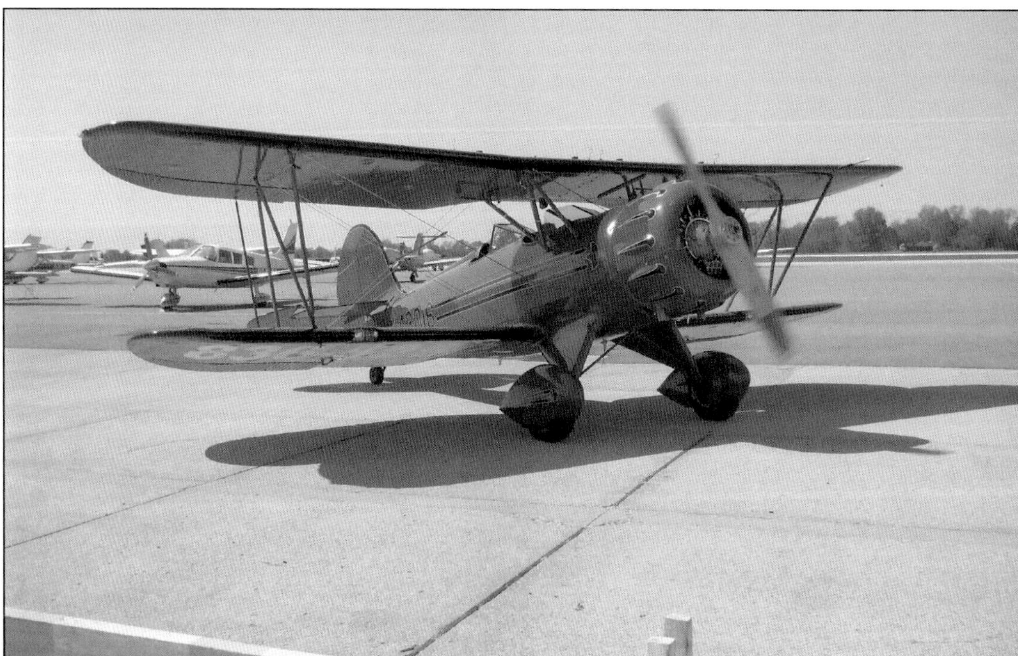

Classic Biplane Tours is the operator of the red WACO biplane that has been a familiar sight over Louisville since 2004. Owner Steve Koch flies the WACO from the back seat, which allows passengers to sit in the front seat of this open cockpit airplane. He has been flying in the Louisville area since 1970 and operates his tour year round, weather permitting. Headquarters for Classic Biplane Tours is the big east side hangar at Bowman Field. (Author's collection.)

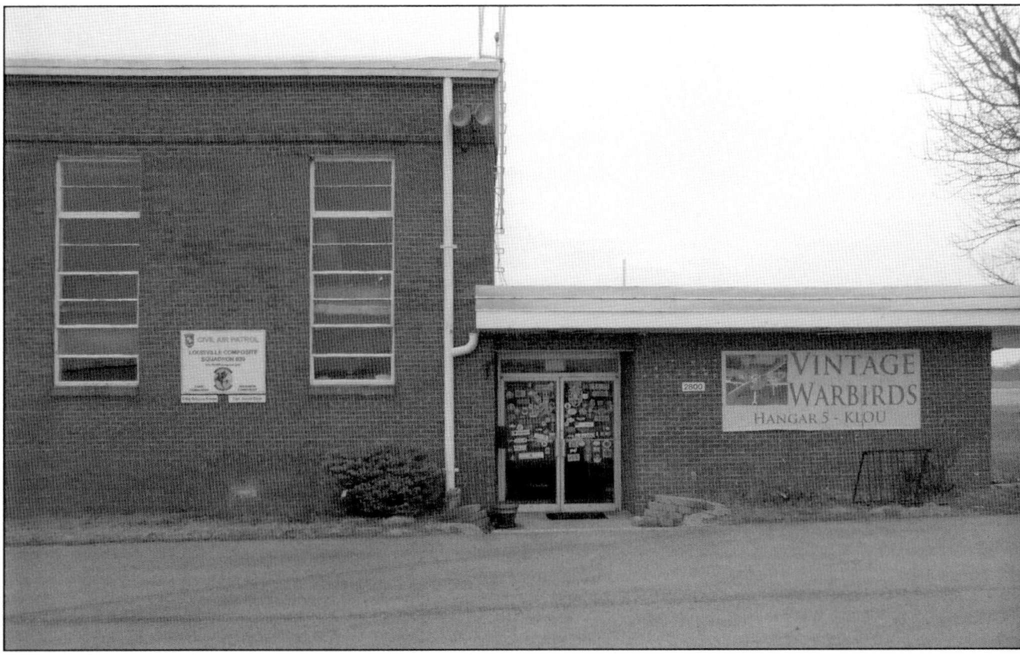

Hangar 5 was one of the last large hangars built at Bowman Field. It was constructed during the 1960s and first opened as Cardinal Aviation. It was used later by Dick Mulloy for his Helicopter Inc. operation and also by Cardinal Wings. Currently, it is the Vintage Warbirds hangar and is maintained by a group of retired military pilots who hangar a selection of vintage propeller-driven military trainers of various manufacturers. Vintage Warbirds schedules events for veterans and flies formation for the annual Thunder Over Louisville air show. The distinctive sound of this group's colorful trainers has become quite common in the skies over Louisville. (Both, author's collection.)

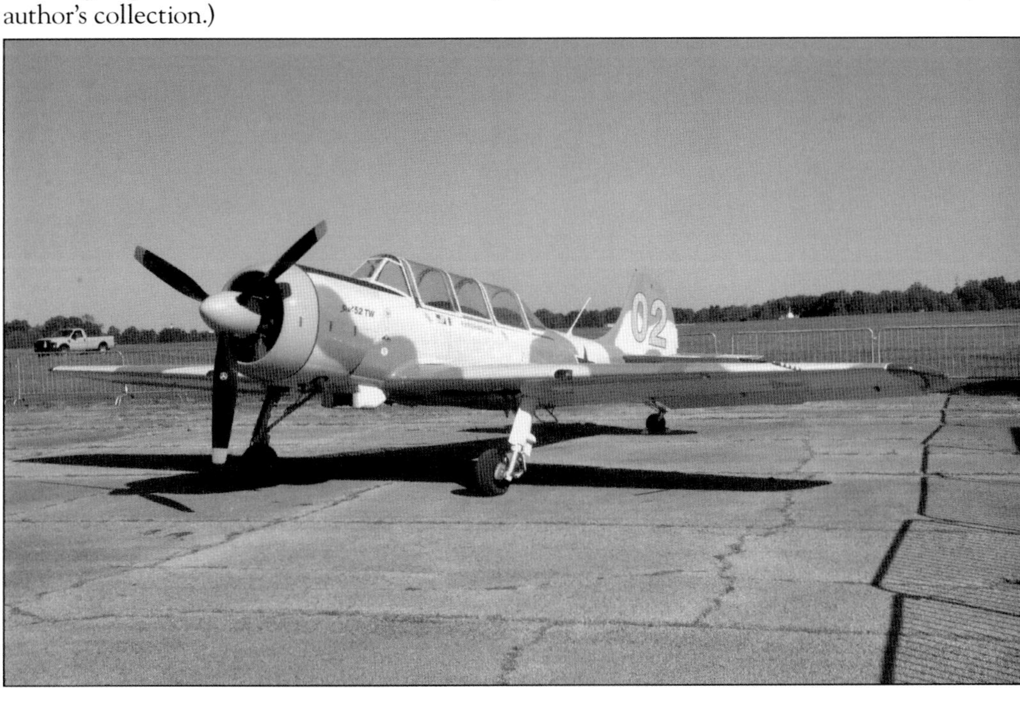

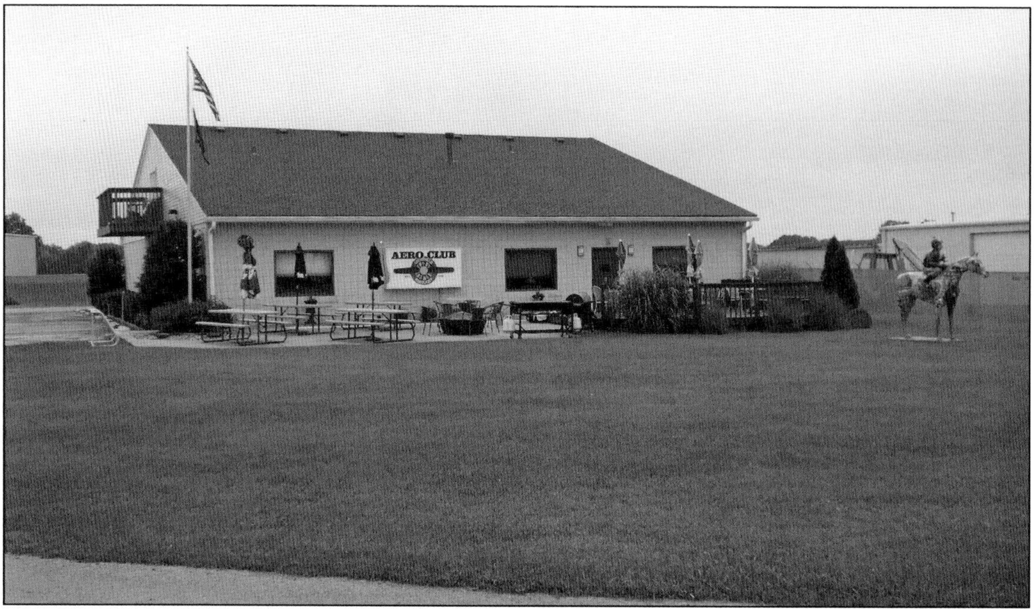

The Aero Club of Louisville has been an important part of Bowman Field since 1922. It has been active in promoting Bowman Field and Louisville aviation and, over the years, has hosted air fairs, dawn patrols, and airport improvement projects. The club has met at various locations around the airport, with its current clubhouse on Bowman Field's east side. The clubhouse's large meeting room is available for rent and is utilized by the Louisville EAA chapter, radio-control model airplane club, and other organizations. (Author's collection.)

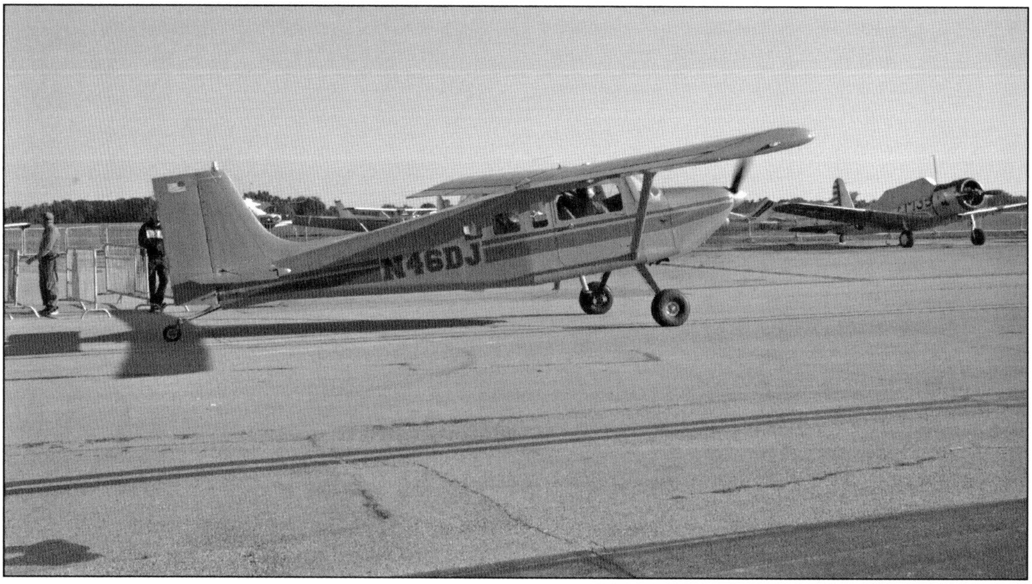

Chapter 110 of the Experimental Aircraft Association, which meets at Bowman Field, dates to 1961 when Howard Edwards petitioned the national EAA headquartered in Wisconsin for information on starting a Louisville chapter. Since then, Chapter 110 has been a leader in promoting aviation to youth through the Young Eagles program, emphasizing aviation safety, and building airplanes from a kit or from scratch. This photograph shows a Murphy SR 2500 built by member Don Jeffries as it arrives at the 2016 Bowman Field Aviation Heritage Festival. (Author's collection.)

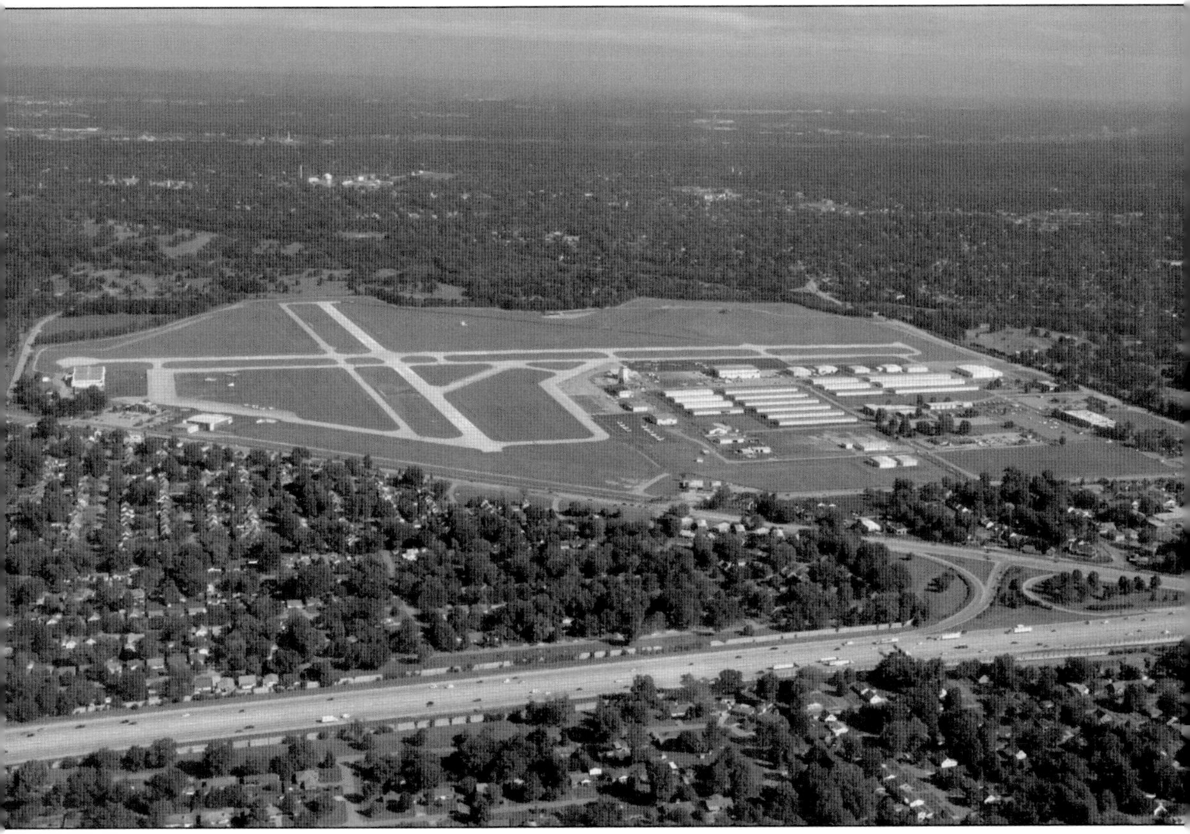

A hundred years later, Bowman Field is clearly seen in this aerial photograph. Many things have changed, especially the airport being surrounded by civilization. Neighborhoods are nearby, and not all airport land is used for aviation, with the Army National Guard and Army Reserve using a portion on the east side along with a driver's license center and the Kentucky Department of Fish & Wildlife. Still, much is the same. Bowman Field continues to be served by excellent road connections that make it easy to get to and from the airport. Just as Bowman Field had all the right things going for it in the beginning, it still does today. (Gary Chambers.)

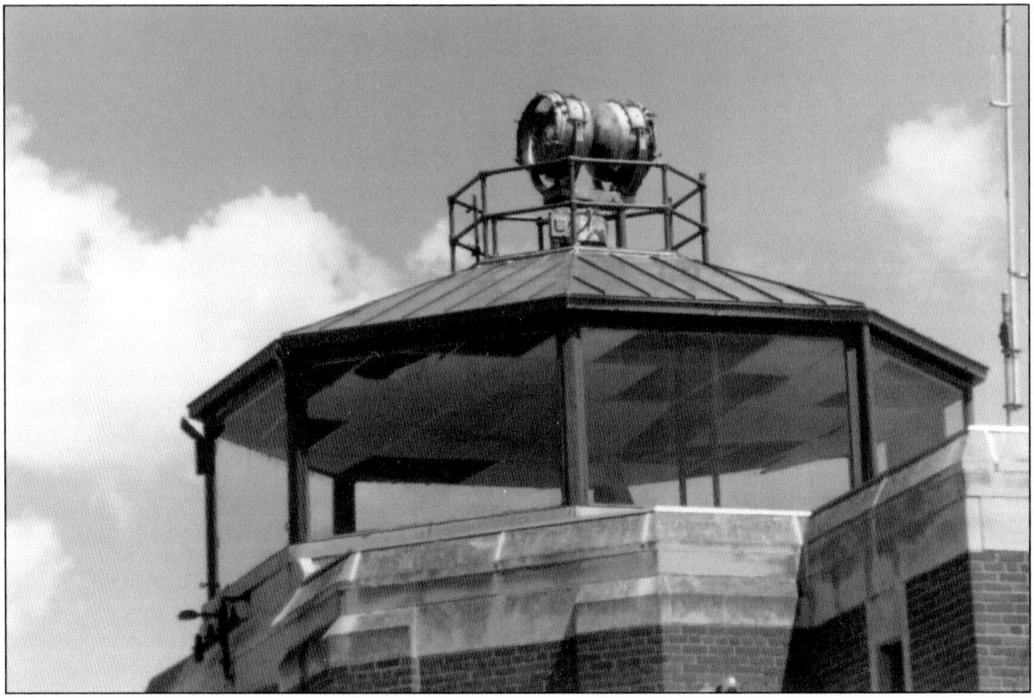

One can only imagine how many airplanes owe a safe landing during the night at Bowman Field to a beam of light that reached out from the rotating beacon light when it was operating atop the Bowman Field administration building. Think about an airliner low on fuel trying to find the airport on a dark night and then spotting the beacon light, which guides it in for a landing. An operating beacon light was in use on top of the administration building from the time of the building's construction until it was replaced in 1991 by a smaller light mounted on a tower, which was located on the east side of the airfield. The non-functioning light is still on the administration building and the new one is still being used. (Both, author's collection.)

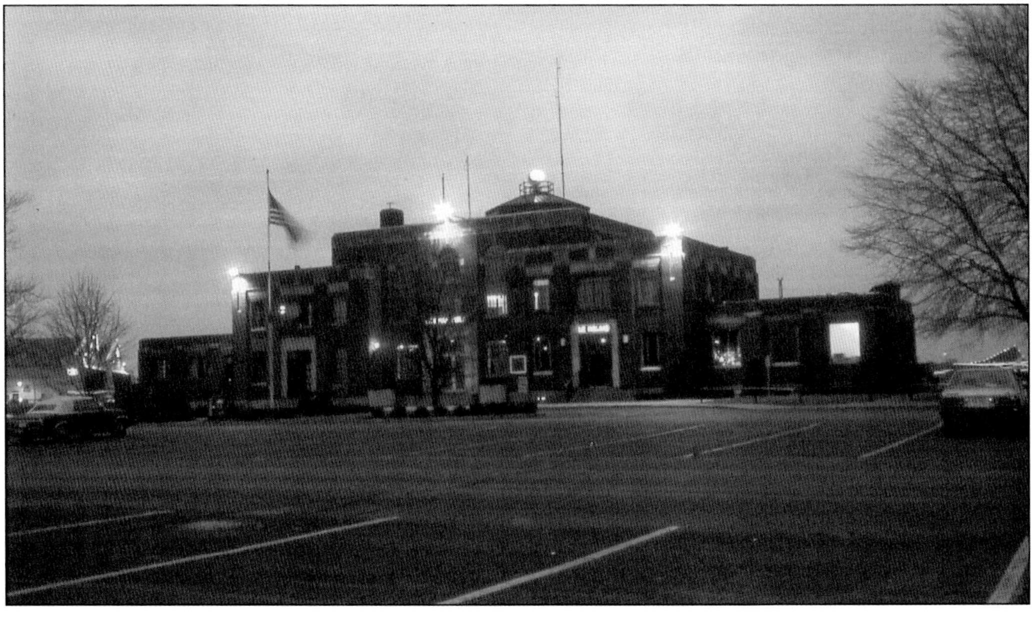

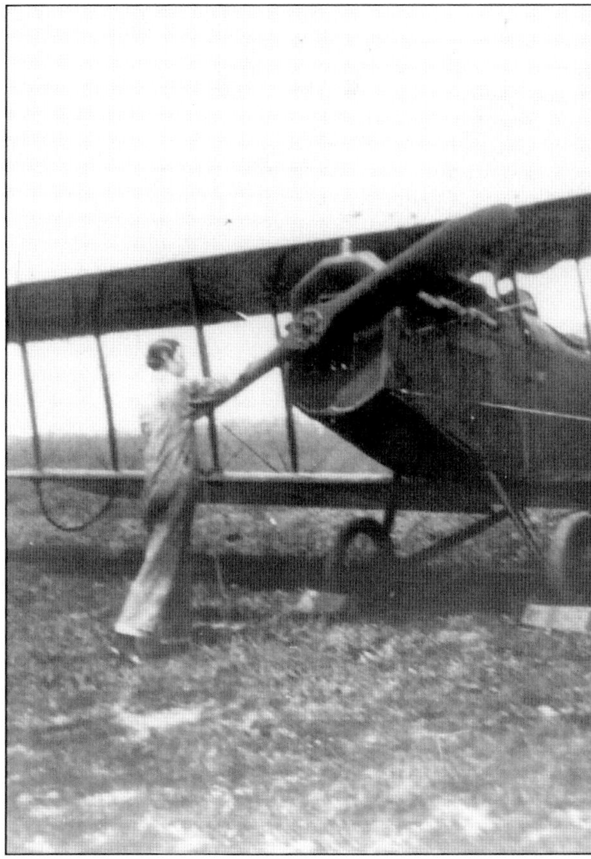

This book began with a photograph of a Curtiss Jenny airplane at Bowman Field in 1920 and ends with an image of a Jenny at Bowman Field in 2016. Although thousands of Curtiss Jennies were made during World War I, the airplane is extremely rare today, especially an airworthy one. Fortunately for Bowman Field, a flying Jenny is located in Bowling Green, Kentucky. This is actually a replica, although it looks and flies like the original. Owner Dorian Walker flies his Jenny on an educational mission to reflect on the important role that the Jenny played in early aviation. And what a beautiful sight it is at 21st-century Bowman Field. (Right, Edward Peck; below, author's collection.)

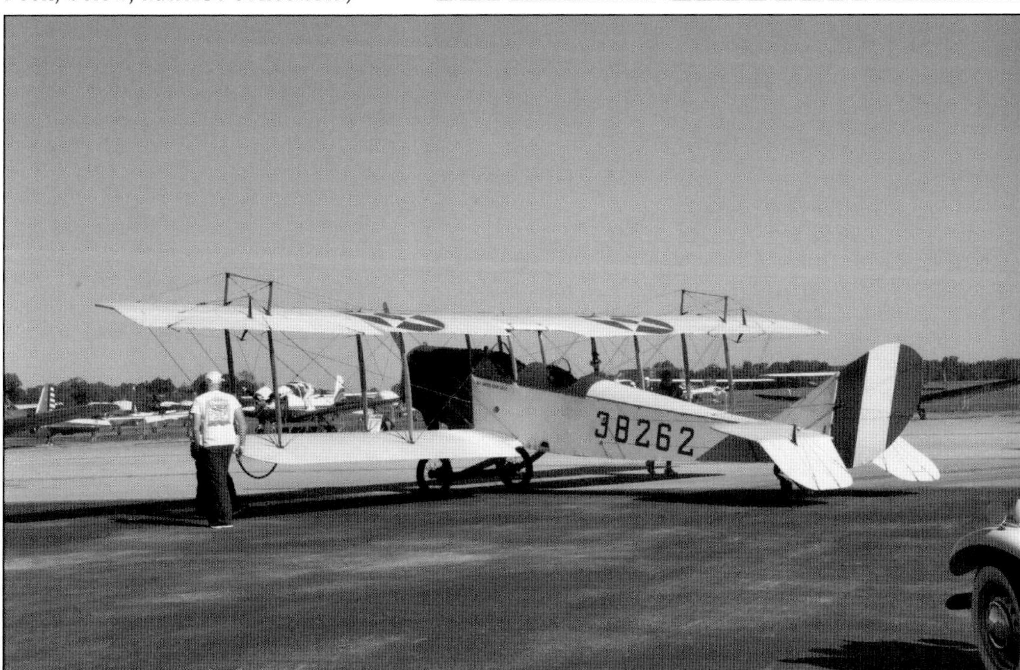

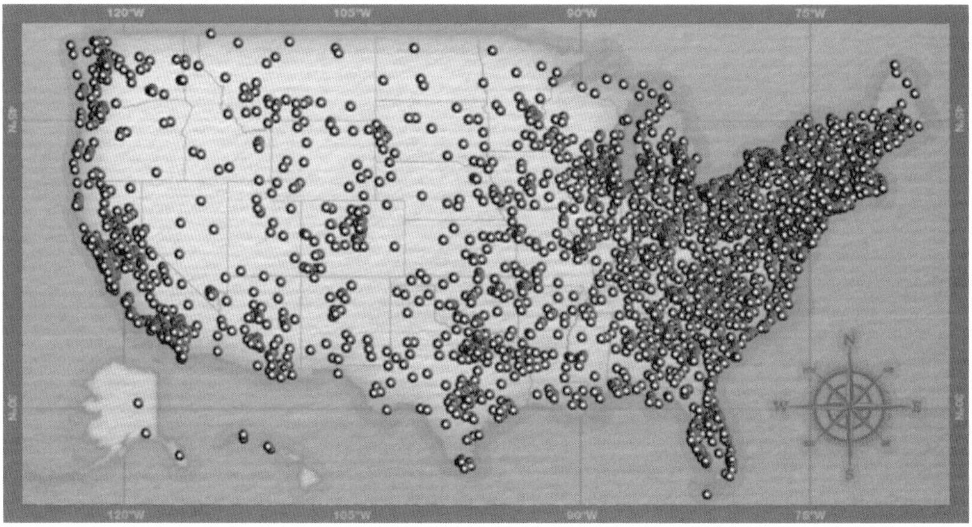